THE RISE OF GAMES AND HIGH-PERFORMANCE COMPUTING FOR
MODELING AND SIMULATION

Committee on Modeling, Simulation, and Games

Standing Committee on Technology Insight—Gauge, Evaluate, and Review

Division on Engineering and Physical Sciences

NATIONAL RESEARCH COUNCIL
OF THE NATIONAL ACADEMIES

THE NATIONAL ACADEMIES PRESS
Washington, D.C.
www.nap.edu

THE NATIONAL ACADEMIES PRESS   500 Fifth Street, N.W.   Washington, DC 20001

NOTICE: The project that is the subject of this report was approved by the Governing Board of the National Research Council, whose members are drawn from the councils of the National Academy of Sciences, the National Academy of Engineering, and the Institute of Medicine. The members of the committee responsible for the report were chosen for their special competences and with regard for appropriate balance.

This is a report of work supported by contract HHM40205D0011, D0#10 between the Defense Intelligence Agency and the National Academy of Sciences. Any opinions, findings, conclusions, or recommendations expressed in this publication are those of the author(s) and do not necessarily reflect the views of the organizations or agencies that provided support for the project.

International Standard Book Number-13: 978-0-309-14777-4
International Standard Book Number-10: 0-309-14777-8

Limited copies are available from:

Division on Engineering and Physical
   Sciences
National Research Council
500 Fifth Street, N.W.
Washington, DC 20001
(202) 334-3118

Additional copies are available from:

The National Academies Press
500 Fifth Street, N.W.
Lockbox 285
Washington, DC 20055
(800) 624-6242 or (202) 334-3313
(in the Washington metropolitan area)
http://www.nap.edu

Copyright 2010 by the National Academy of Sciences. All rights reserved.

Printed in the United States of America

THE NATIONAL ACADEMIES
Advisers to the Nation on Science, Engineering, and Medicine

The **National Academy of Sciences** is a private, nonprofit, self-perpetuating society of distinguished scholars engaged in scientific and engineering research, dedicated to the furtherance of science and technology and to their use for the general welfare. Upon the authority of the charter granted to it by the Congress in 1863, the Academy has a mandate that requires it to advise the federal government on scientific and technical matters. Dr. Ralph J. Cicerone is president of the National Academy of Sciences.

The **National Academy of Engineering** was established in 1964, under the charter of the National Academy of Sciences, as a parallel organization of outstanding engineers. It is autonomous in its administration and in the selection of its members, sharing with the National Academy of Sciences the responsibility for advising the federal government. The National Academy of Engineering also sponsors engineering programs aimed at meeting national needs, encourages education and research, and recognizes the superior achievements of engineers. Dr. Charles M. Vest is president of the National Academy of Engineering.

The **Institute of Medicine** was established in 1970 by the National Academy of Sciences to secure the services of eminent members of appropriate professions in the examination of policy matters pertaining to the health of the public. The Institute acts under the responsibility given to the National Academy of Sciences by its congressional charter to be an adviser to the federal government and, upon its own initiative, to identify issues of medical care, research, and education. Dr. Harvey V. Fineberg is president of the Institute of Medicine.

The **National Research Council** was organized by the National Academy of Sciences in 1916 to associate the broad community of science and technology with the Academy's purposes of furthering knowledge and advising the federal government. Functioning in accordance with general policies determined by the Academy, the Council has become the principal operating agency of both the National Academy of Sciences and the National Academy of Engineering in providing services to the government, the public, and the scientific and engineering communities. The Council is administered jointly by both Academies and the Institute of Medicine. Dr. Ralph J. Cicerone and Dr. Charles M. Vest are chair and vice chair, respectively, of the National Research Council.

www.national-academies.org

COMMITTEE ON MODELING, SIMULATION, AND GAMES

DELORES M. ETTER, *Co-Chair*, Southern Methodist University
ROBERT J. HERMANN, *Co-Chair*, Global Technology Partners, LLC
BRIAN BALLARD, ARES Systems Group
PIERRE CHAO, Renaissance Strategic Advisors
ROBERT GEHORSAM, Forterra Systems, Inc.
SHARON C. GLOTZER, University of Michigan
J. C. HERZ, Batchtags, LLC
ALLISON A. HICKEY, Accenture National Security Services
CHARLES HUDSON, Serious Business
JAMES PEERY, Sandia National Laboratories
BENJAMIN SAWYER, Digitalmill, Inc.
ETHAN WATRALL, Michigan State University
MICHAEL J. ZYDA, University of Southern California

Staff

MICHAEL A. CLARKE, Lead Board Director
DANIEL TALMAGE, Study Director
CARTER FORD, Program Officer
LISA COCKRELL, Associate Program Officer
ERIN FITZGERALD, Associate Program Officer
SARAH CAPOTE, Research Associate
SHANNON THOMAS, Program Associate

Preface

Rapid expansion in computer processing, storage, communications, and display capability has resulted in the proliferation of new software for modeling, simulation, and games. Computational models have the capability to produce increasingly useful simulations to predict natural phenomena and engineered systems, study human behavior and physiology, or educate. An increased cultural acceptance of electronic games for a wide variety of applications—including entertainment, education, training, and rehabilitation—has elevated the video games industry into a position of unprecedented significance and legitimacy as a media form. In the meantime, the increased ubiquity of computing power has lowered the threshold for obtaining effective modeling and simulation capabilities. The globalization of technology and increased affordability of hardware have created an environment in which major developments can originate not only from recognized industry and academic sources in the United States but also from individuals in virtually any location in the world.

In spring 2008 the Defense Intelligence Agency (DIA) asked the National Research Council (NRC) to form a committee to investigate the possible significance to national security of new modeling, simulation, and games technologies. In this report the Committee on Modeling, Simulation, and Games assesses the current state of modeling, simulation, and games research and development worldwide. The committee also identifies ways in which these technologies could impact government and society in the future.

We wish to express our sincere appreciation to the members of the committee for their contributions to the preparation of this report and to the staff of DIA for their sponsorship. On behalf of the entire committee, we also thank NRC staff members Michael A. Clarke, Daniel Talmage, Erin Fitzgerald, Carter Ford, Lisa Cockrell, Sarah Capote, and Shannon Thomas for their support and assistance in the production of this report.

Delores M. Etter
Robert J. Hermann
Co-Chairs
Committee on Modeling, Simulation, and Games

Acknowledgment of Reviewers

This report has been reviewed in draft form by individuals chosen for their diverse perspectives and technical expertise, in accordance with procedures approved by the National Research Council's Report Review Committee. The purpose of this independent review is to provide candid and critical comments that will assist the institution in making its published report as sound as possible and to ensure that the report meets institutional standards for objectivity, evidence, and responsiveness to the study charge. The review comments and draft manuscript remain confidential to protect the integrity of the deliberative process. We wish to thank the following individuals for their review of this report:

Steve Arnold, Polaris Ventures
R. Stephen Berry, NAS, University of Chicago
W. Peter Cherry, NAE, Science Applications International Corporation
David B. Kirk, NAE, NVIDIA
Gilman Louie, Alsop Louie Partners
Marisa Ong, Total Immersion Software

Although the reviewers listed above have provided many constructive comments and suggestions, they were not asked to endorse the conclusions or recommendations, nor did they see the final draft of the report before its release. The review of this report was overseen by George Hornberger, NAE, Vanderbilt University, and Frank E. Brown, Raytheon Corporation. Appointed by the NRC, they were responsible for making certain that an independent examination of this report was carried out in accordance with institutional procedures and that all review comments were carefully considered. Responsibility for the final content of this report rests entirely with the authoring committee and the institution.

Contents

SUMMARY ... 1

1 A NEW PARADIGM IN MODELING AND SIMULATION ... 7
   Introduction and Study Origin, 7
   Motivation, 8
   Structure of This Report, 9
   References, 9

2 MODELING, SIMULATION, GAMES, AND COMPUTING ... 10
   Introduction, 10
   The Path to Exascale Computing, 11
      The Evolution of Computing Architectures: From ENIAC to Multicore, 11
      Multicore Processing, 12
      CPU Accelerators: Graphical Processing Units, 14
      CPU Accelerators: Field-Programmable Gate Arrays, 17
      Roadmap for Future High-Performance Computing, 17
      Software for Exascale Computing, 19
      Future Technologies Enabled by Exascale Computing, 20
   Human Capital in Science-Based Modeling and Simulation, 21
   Human Capital in Computer and Video Games, 23
      International Presence, 25
      Industry Implications of Formal Academic Game Programs, 26
   Final Thoughts, 27
   References, 27
      Published, 27
      Unpublished, 29

3 GAMES: BEYOND ENTERTAINMENT   30
   Introduction, 30
   Games and Culture, 31
      Interactive and Participatory Culture, 31
      Transmedia and Popular Culture, 33
      Political and Other Simulation Games, 34
   Outputs and Effects of Game Play, 35
      Context of Game Play, 35
      Understanding the Effects of Game Play, 36
      Democratization, 39
   Serious Games, 40
      Cultural Attitudes Toward Serious Games, 41
      Educational and Training Dimensions of Serious Games, 42
   The Business of Games, 42
      Evolution of the Games Industry, 43
      Global Industry Trends, 44
      International Competition in Game Development, 48
   Final Thoughts, 51
   References, 52
      Published, 52
      Unpublished, 54

4 DEFENSE MODELING, SIMULATION, AND GAMES   55
   Introduction, 55
   Scientific Modeling and Simulation, 56
   Cyber and Kinetic Warfare, 59
      Computer Security, 60
   Cyber Propaganda Through Games, 60
      Political Manipulation Through Games on the Internet, 61
   War Games, 61
      The Evolution of War Games, 61
      Major Applications of Military War Games, 63
      Enhanced Military Simulation, 67
   Final Thoughts, 70
      Take-away Warnings, 70
      Take-away Opportunities, 71
   References, 71
      Published, 71
      Unpublished, 72

APPENDIXES

A	Biographical Sketches of Committee Members	75
B	Meetings and Speakers	81
C	Committee Methodology	85
D	Key Recommendations from Previous Studies	92
E	An Overview of Digital Games	98

Acronyms and Abbreviations

AI	artificial intelligence
AMD	Advanced Micro Devices
API	application programming interface
ARG	alternate-reality games
ASC	Advanced Simulation and Computing program
BMI	body mass index
CONOPs	concept of operations
COTS	commercial off-the-shelf
CPU	central processing unit
CRPG	computer role-playing games
CUDA	Compute Unified Device Architecture
DARPA	Defense Advanced Research Projects Agency
DIA	Defense Intelligence Agency
DoD	Department of Defense
DS	dual screen
EFP	Explosively Fired Projectile
ENIAC	Electronic Numerical Integrator and Computer
FLOPS	floating-point operations per second
FPGA	field-programmable gate array
FPS	first-person shooter (game type)

GPU	graphical processing unit
HPC	high-performance computing
IC	intelligence community
IGDA	Independent Game Developers Association
IP	intellectual property
IP	Internet Protocol
LAN	large area network
M&S	modeling and simulation
MMOG	massively multiplayer online game
MMORPG	massively multiplayer online role-playing game
MS&G	modeling, simulation, and games
MUD	*Multi-User Dungeon*
NNSA	National Nuclear Security Administration
NRC	National Research Council
NSF	National Science Foundation
OS	operating system
PC	personal computer
PS3	Sony PlayStation 3
PSP	PlayStation Portable
PSYOPs	psychological operations
R&D	research and development
RPG	role-playing game
RTS	real-time strategy (game)
SIMD	single instruction, multiple data
TBS	turn-based strategy (game)
TIGER	Technology Insight—Gauge, Evaluate, and Review
TTPs	tactics, techniques, and procedures
UQ	uncertainty quantification
USC	University of Southern California
V&V	verification and validation
WWDC	Apple Worldwide Developers Conference
WoW	*World of Warcraft*

Summary

The technical and cultural boundaries between modeling, simulation, and games are increasingly blurring, providing broader access to capabilities in modeling and simulation and further credibility to game-based applications. The purpose of this study, carried out by the National Research Council's Committee on Modeling, Simulation, and Games, is to provide a technical assessment of modeling, simulation, and games (MS&G) research and development worldwide and to identify future applications of this technology and its potential impacts on government and society. Further, this study identifies feasible applications of gaming and simulation for military systems; associated vulnerabilities of, risks to, and impacts on critical defense capabilities; and other significant indicators and warnings that can help prevent or mitigate surprises related to technology applications by those with hostile intent. Finally, this report recommends priorities for future action by appropriate departments of the intelligence community (IC),[1] the Department of Defense (DoD) research community, and other government entities.

It is the intention of the committee that the results of this study serve as a useful tutorial and reference document for this particular era in the evolution of MS&G. The report also highlights a number of rising capabilities to watch for that are facilitated by MS&G.

MODELING, SIMULATION, GAMES, AND COMPUTING

Advances in hardware and software for computation provide an essential basis for improving modeling and simulation. Supercomputing performance has increased by 14 orders of magnitude in the past 60 years. The most dramatic increase has occurred over the past 20 years, with the advent of massively parallel computers and associated programming paradigms and algorithms. Moreover, recent years have seen the commoditization of many of these technologies. Instead of the very expensive, special-purpose hardware found in vector platforms, commercial off-the-shelf parts can now be connected with networks to create supercomputers.

---

[1]The IC is made up of approximately 18 entities across the executive branch. A detailed listing of all IC members is found on the U.S. Intelligence Community Web site at http://www.intelligence.gov/1-members.shtml. Last accessed on June 24, 2009.

However, the computing future presents new challenges. As component sizes continue to shrink and the processing speed of central processing units (CPUs) has plateaued, CPU vendors are adding additional processing units or "cores" to a single chip (a form of parallelism) to achieve performance gains that will present significant challenges to the development of effective and efficient scalable software on a node. Next-generation supercomputers will rely on many thousands of multicore nodes working together, presenting numerous challenges in the areas of resiliency, power, cooling, and scalability.

Multicore processors and accelerators, including inexpensive and abundant graphical processing units, are changing the landscape of computing through a new era of on-chip parallelism. Also, programming models, libraries and compilers that automate parallelization and threading of sequential code, and a new generation of computer programmers who think and code in a multithreaded parallel way, would open these future high-performance computing (HPC) platforms to a broad set of application developers.

Once the purview of technologically sophisticated and wealthy nations, modeling and simulation capabilities are now well within the reach of most state and nonstate actors. Advances in computation, representational algorithms, and the software to harness these advances will continue and will be accessible to all developed and developing societies. However, the increased ubiquity of computing power has lowered the threshold for obtaining effective modeling and simulation capabilities. No state or nonstate actor is likely to maintain a sustainable, strategic, comparative advantage in access to these capabilities. From a national security and an HPC leadership perspective, there will be international competition to develop the capacity for usable exascale[2] computing, with specific "breakthrough" points to be achieved.

The scientific and systems skills needed to create algorithms that capture the underlying scientific and system phenomena to be represented by models and simulations are important elements of capability. There is a notable lack of highly skilled computational scientists and engineers able to fully leverage the current state of the art in HPC for science-based modeling and simulation (NRC, 2007; WTEC, 2009). New computing architectures for games will continue to lead to new capabilities provided by modeling and simulation, including increased predictive accuracy. As computers have become faster, models have increased in fidelity, causing games to become more realistic and accurate. Second, the increasing fidelity of games has led game makers to seek the manufacture of higher-performance computer chips. As such, the need for a skilled workforce in modeling and simulation, and in simulation-based engineering and science generally, will become increasingly important to national security to take advantage of improvements in computing speed and accuracy.[3]

GAMES: BEYOND ENTERTAINMENT

The games industry has evolved over time from purely commercial entertainment to more recent applications of games by civil, commercial, and military organizations (WTEC, 2009). The games industry and the personal computers industry have provided innovations in graphics and PC hardware that have led, for example, to PC-based simulations. Until relatively recently, most military and government simulations were performed on workstations (i.e., SGI, Sun) and on specially designed hardware (i.e., Evans & Sutherland, Delta Graphics).

Evidence indicates that the use of games might have substantial impacts on human and group behav-

---

[2]Systems that can handle a million trillion, or 10^{18}, floating-point calculations per second.

[3]The committee notes that a good reference for simulation-based engineering is the National Science Foundation report *Simulation-Based Engineering Science*, a summary of which is provided online at http://www.nsf.gov/attachments/106803/public/TO_SBES_Debrief_050306.pdf. Last accessed on October 14, 2009.

ior, encouraging skill development and facilitating new approaches to how to communicate, organize, and act in ad hoc collaborative environments. However, given the early state of growth of these "serious" applications of games, sufficient research has not been done to permit definitive conclusions about the scope and nature of these potential impacts. Measuring the emotional, psychological, and societal effects of broad applications of games may never be precise, but the potential of games is clearly worthy of consideration by DoD as both a tactical advantage and a tactical threat.

The committee has identified many global industry trends in games and included them in this report, including:

- The rise of massively multiplayer online games;
- The intersection of productivity and games (e.g., prediction markets);
- The intersection of social networks and games;
- Mobile games platforms;
- The microtransaction-based business model; and
- Highly realistic three-dimensional modeling and representation of Earth and real geographic locations (e.g., Google Earth, DARPA's RealWorld, Microsoft's Virtual Earth, first-person shooters).

Beyond these, there has been a notable increase in the development of *serious games*, which have a purpose beyond entertainment, such as skill development or improved physiological and psychological health. They can be original programs created for a specific utility or programs repurposed from existing entertainment-focused games. The acceleration of serious gaming is fueled not only by the gains made in video game design, audience reach, and technology but also by the underlying advancement of core technologies such as the Internet and social networking, and the gains described by Moore's law and Metcalfe's law. The progression of game design and game capabilities as a result has yielded increased interest in applying those capabilities in support of nonentertainment purposes.

APPLICATIONS TO DEFENSE

Complex networked systems of systems can be effectively designed and tested only with the help of modeling and simulation (Brase and Brown, 2009). Adding games to this set of capabilities permits better modeling and understanding of human behavior as well. As the two have become connected, games have taken on new relevance for defense analysis. Modeling, simulation, and games relating to complex systems will become increasingly important to the United States and its adversaries.

FINDINGS AND RECOMMENDATIONS

The findings and recommendations given below emphasize that modeling, simulation, and games are evolving rapidly and merit attention by the IC.

Finding 2-1: "Cache-friendly" algorithms have been developed, but many take longer to run than "non-friendly" algorithms. For intelligence analysts responsible for technology warning, a major breakthrough in memory speeds would be a game changer and have significant national security implications. One area showing promise is three-dimensional packaging of the memory and CPU, which provides for more pins, and thus higher bandwidth, to be connected between the two devices (Kogge et al., 2008). Rather than monitoring advances in processor speeds, tracking improvements in memory speed could provide earlier warning of the next step change in capabilities.

Finding 2-2: Exaflop-level computing is expected to be aggressively pursued by the United States, Europe, and Asia (WTEC, 2009). Areas for intelligence analysts responsible for technology warning to watch that could facilitate a breakthrough for usable exaflop computing include technologies that:

- Significantly reduce memory power consumption;
- Dramatically increase memory-to-processor communication speed while reducing latency;
- Dramatically increase processor-to-processor communication speed while reducing latency;
- Dramatically reduce part counts through integration or packaging that will increase the mean time to failure; or
- Automatically generate scalable code.

Finding 3-1: As game development further becomes its own formal discipline taught in universities and is merged with existing modeling and simulation programs, the result will be a generation of practitioners more apt to apply innovations from game design and development to the larger modeling and simulation community.

Finding 4-1: Improvements in and the deployment of agent-based simulation technology—that is, technology that simulates the actions and interactions of autonomous characters and/or systems such that an understanding or a view into the simulated behavior or system can be obtained—as the underpinning of game artificial intelligence systems could be a source of significant vulnerability to the extent that the United States falls behind in this area.

Agent-based simulation technology provides a computational analytic framework by allowing the exploration of potential outcomes such that an analysis can be performed for a system for which no easy, closed-form analytic solution is possible.

Recommendation 4-1: Military war games should exploit the significant growth and lessons of serious games to leverage experiential aspects of large multiplayer joint war games. A more real-time, immediate-feedback exploration environment can then be assessed using rapidly updated algorithms, parameters, and coefficients that reflect behavioral and policy implications. The use of serious games to explore strategy or technology implications can be valuable to strategic and long-range concepts of operations, weapons system acquisition, and threat assessment and response and far more effective in providing constant assessment of new technology and CONOPs opportunities, as well as more real-time threat warnings to those who consistently monitor for these issues. This same virtual sandbox can provide rapid assessment of these capabilities in a more affordable virtual method and reduce the manpower-extensive planning, logistical, and cost requirements typically required by large war games that only occur on a two-year basis.

Finding 4-2: While the United States continues to leverage superior training as part of its ability to maintain asymmetric advantages over potential adversaries, these same potential adversaries may develop the ability to train and adapt CONOPs based on prolific access to Western game genres and actors.

Finding 4-3: In the global war on terrorism, American forces have frequently prevailed in direct-fire exchanges, often attributed to better squad coordination and training—a skill commercial multiplayer gamers practice and develop virtually and routinely. Though it is not expected that this advantage will be maintained when engaged with a sophisticated potential adversary in the field, the United States is at risk of losing its advantage due to the advanced training environments and distributed nature of simulation and online gaming available either currently or in the near future.

Recommendation 4-2: DoD should strongly consider migrating at least one of its Title 10 war games to the emerging architectures of the commercial gaming industry.

The computational power in mobile devices is rapidly increasing due in part to the popularity of mobile gaming. Additional increases in cellular infrastructure and cellular network speeds are happening at a faster clip outside the United States. This country may soon find itself living with an inadequate cellular infrastructure due to events taking place in the commercial cellular market. Aging and outdated digital infrastructure is already showing signs of hindering progress. At the 2009 Apple Worldwide Developers Conference, the U.S.-based company announced multiple features and capabilities for the iPhone, such as Multimedia Messaging Services and tethering, that will appear *outside* the United States (on 22 major carriers), only to *possibly* later appear within the United States if the U.S. carrier allows making those features possible. The delay, based predominantly on concerns about supporting increased bandwidth requirements and billing plans, underscores unpreparedness for dealing with a rapidly changing technology and the increased pervasiveness of ubiquitous mobile computing. Additional technologies and capabilities thought to have the potential to greatly impact MS&G and related fields are called out in Box S-1 and expanded on via charts in this report.

Simulations in the context of computing are often thought of as running an executable to compute something on a macro or nano level. In contrast, games are meant to involve the player and to incorporate human behavior in a system. Using modeling, simulation, and games in tandem may facilitate better innovation, better accessibility to actors around the world, and better models of human system dynamics for commercial and defense systems.

BOX S-1
Technology Forecasting Methodology in This Report

The report *Avoiding Surprise in an Era of Global Technology Advances* (NRC, 2005) identifies a methodology for the intelligence community (IC) that has been widely accepted as a tool for finding and recognizing potential future national security threats from different emerging technologies. This methodology, described further in Appendix C, provides the IC with a means to gauge the potential implications of emerging technologies.

Specific technology topics addressed in the current report via the technology forecasting methodology include:

Chapter 2:
- Collapsing the memory wall (a counterforce to Moore's law)
- Software for massively parallel architectures

Chapter 4:
- Automation of verification, validation, and uncertainty quantification
- Vulnerability of physical assets given open-source software and information
- Reverse engineering software
- Cyber warfare through virtual worlds
- Presimulated automated computer attacks
- Behavioral modeling and political manipulation through virtual environments

REFERENCES

Brase, James M., and David L. Brown. 2009. *Modeling, Simulation and Analysis of Complex Networked Systems: A Program Plan*. Livermore, CA: Lawrence Livermore National Laboratory. Available at https://wiki.cac.washington.edu/download/attachments/7478403/ComplexNetworkedSystemsProgram-final.pdf?version=1. Accessed November 12, 2009.

Kogge, Peter M., Keren Bergman, Shekhar Borkar, Dan Campbell, William Carlson, William Dally, Monty Denneau, Paul Franzon, William Harrod, Kerry Hill, Jon Hiller, Sherman Karp, Stephen Keckler, Dean Klein, Robert Lucas, Mark Richards, Al Scarpelli, Steven Scott, Allan Snavely, Thomas Sterling, R. Stanley Williams, and Katherine Yelick. 2008. *Exascale Computing Study: Technology Challenges in Achieving Exascale Systems*. Atlanta: Georgia Institute of Technology. Available at http://users.ece.gatech.edu/~mrichard/ExascaleComputingStudyReports/ECS_reports.htm. Accessed June 22, 2009.

NRC (National Research Council). 2005. *Avoiding Surprise in an Era of Global Technology Advances*. Washington, DC: The National Academies Press. Available from http://www.nap.edu/catalog.php?record_id=11286.

NRC. 2007. *Rising Above the Gathering Storm: Energizing and Employing America for a Brighter Economic Future*. Washington, DC: The National Academies Press. Available at http://www.nap.edu/catalog.php?record_id=11463.

WTEC (World Technology Evaluation Center). 2009. Sharon C. Glotzer, Sangtae Kim, Peter T. Cummings, Abhijit Deshmukh, Martin Head-Gordon, George Karniadakis, Linda Petzold, Celeste Sagui, and Masanobu Shinozuka, panel members. *International Assessment of Simulation-Based Engineering and Science*. Baltimore, MD: WTEC. Available at http://www.wtec.org/sbes. Accessed June 22, 2009.

1

A New Paradigm in Modeling and Simulation

INTRODUCTION AND STUDY ORIGIN

This is the fifth report in a series produced under the support of the National Research Council (NRC) Standing Committee on Technology Insight—Gauge, Evaluate, and Review (TIGER) and sponsored by the Defense Intelligence Agency's Defense Warning Office (NRC, 2006, 2008a, 2008b, 2009). As with the previous reports, sponsorship of the current report was a direct result of discussions between the standing committee and the U.S. intelligence community (IC). The overall series is intended to help the IC ascertain global technology trends that could potentially affect future U.S. warfighting capabilities.

An earlier report, *Avoiding Surprise in an Era of Global Technology Advances* (NRC, 2005), produced a methodology for technology warning that previously had not been available to the IC and since then has been accepted by the IC as a tool for finding and recognizing potential national security threats stemming from emerging technologies. This methodology, described further in Appendix C, provides the IC with a means to gauge the potential implications of emerging technologies. As part of a continuing relationship with the TIGER standing committee, the IC identified modeling, simulation, and games (MS&G) research and development as a field that could pose strategic implications for U.S. national security. Box 1-1 provides the study's statement of task.

Recent decades have yielded rapid progress in the technical capabilities and ubiquity of MS&G. The Committee on Modeling, Simulation, and Games[1] was established to survey the technical state of these fields and to evaluate progress made since the publication of related NRC reports on modeling and simulation (summarized in Appendix D). A major new contribution comes from the broadened emphasis on games and their effect on culture and applications as opposed to the roles served by modeling and simulation in the past.

---

[1] A listing of committee members and their biographies can be found in Appendix A.

> **BOX 1-1**
> **Statement of Task**
>
> An Ad Hoc Committee of the National Research Council (NRC) will provide a technical assessment of worldwide modeling, simulation, and games research and development (R&D). The study will outline the current state of the art, and use the methodology presented in the 2005 NRC report *Avoiding Surprise in an Era of Global Technology Advances,* to identify future applications of this technology and its potential impact on government and society.
> Specifically, the committee will:
>
> - Examine how emerging computer architecture, algorithms, and software impact games and engineering simulation.
> - Consider the importance of simulations that test algorithms of automated systems, especially those built as artificial intelligence for weapons systems.
> - Focus on how games affect attitudes and values of the people who play them, including examining how cultural difference causes approaches to gaming to differ.
> - Examine political games, including how players participate in them, including god games (which is a term of art denoting a subgenre of construction and management simulation games) and the ramification on policy and culture; persistent worlds; as well as network games.
> - Identify how simulation and games could be best developed to affect cyber and kinetic warfare efforts, including looking at using gaming in future planning for new weapons systems and platforms.
> - Analyze games in the context of other technologies used by gamers to share information and organize activities, to include: social networking sites, voice over Internet protocol capabilities, mobile devices, and Web 2.0 capabilities.

MOTIVATION

While modeling, simulation, and digital games have all seen rapid progress in recent years with respect to increased fidelity, realism, and sophistication, as well as better human and physical modeling, the motivation for coupling these concepts in this study may not initially be clear to the reader. In fact, not only does each field influence the other but they can serve as complementary approaches for similar end goals. As the committee discussed the distinctions between modeling, simulation, and games, several general observations emerged.

Current modeling and simulation work includes focuses on financial markets, natural processes, population, and food production, as well as nonstatic modeling of materials and physical systems across a wide range of scales, from subatomic to planetary. In contrast, games "model" the world at the human and social levels. Given the commercial nature of games, there is an inherent focus on creating human experiences and evoking responses from the player. Such capabilities are a critical distinction from those achievable in nontraining government simulations. Depending on the genre of the game (for examples, see Appendix E), the behavior modeled may range from the individual level to the tribal or even the nation-state level, at which point economic, military, and physical modeling of low-to-moderate fidelity may come into play.

The essence of games is, of course, the human-in-the-loop, interactive play component. In most simulations, especially military simulations, human participation is not restricted to input and output receipt. In many cases, humans are in multiple loops—"operators" who are in control of the scenario and tasking of automated and semiautomated forces, humans controlling avatars (like soldiers) and vehicles (like helicopters, tanks, trucks, and fighters) in the simulator. Some of these simulators (like the Defense Advanced Research Projects Agency's SIMNET in the 1980s and RealWorld today) are networked and support full communications. Modeling humans is as important in simulations as in massively multi-player games. In addition, while simulations generally require high-performance computing capabilities, platform games typically run on consumer-grade hardware (though online games now require large data centers of commodity servers with high bandwidth to support high levels of concurrency).

While the committee observed that MS&G-related technologies as a whole (both hardware and software) are increasingly accessible to a wide range of global players, it is also true that the global nature of the games industry and its relatively low computing requirements make games potentially more available as a national security resource to actors friendly and unfriendly to the United States. Finally, it is possible to conceive that games as a whole have the potential to fill a gap in current modeling and simulation efforts not likely to be addressed adequately by current Department of Defense efforts in terms of time, quality, and lack of innovation. That gap lies primarily in the applicability of MS&G-related technologies to human activity.

STRUCTURE OF THIS REPORT

This report provides an overview of research in modeling, simulation, and games and poses a series of questions of likely relevance to decision makers.

Chapter 2 provides detail on factors that have led to the recent rise in the usage and utility, including growing computational power and human capital, and factors deemed by the committee to have the potential to transform MS&G in the future.

Chapter 3 emphasizes games specifically. It serves as a tutorial on the relevant components, vocabulary, and recent developments of the field and provides an analysis of the history and potential of games to impact economics, social interaction, and culture.

Chapter 4 applies the take-away messages from the previous two chapters to survey MS&G in a defense context. Through war games, cyber propaganda, and other security issues, it is shown that the United States is in a position to take advantage of some of the exciting new applications in the field of MS&G, but the global investment in these areas (and broadened access to U.S. tools and technologies) means that these technologies must be watched carefully.

Appendix C (Chapter 2 from *Avoiding Surprise*) provides a full account and explanation of the technology warning methodology and terminology used in Chapters 2, 3, and 4. Readers will find it helpful to read Appendix C before reading the rest of this report. Appendix D summarizes key findings from three related modeling and simulation reports, and Appendix E serves as a tutorial on the types, platforms, and business structures of games and the games industry.

REFERENCES

NRC (National Research Council). 2005. *Avoiding Surprise in an Era of Global Technology Advances*. Washington, DC: The National Academies Press. Available from http://www.nap.edu/catalog.php?record_id=11286.

NRC. 2006. *Critical Technology Accessibility*. Washington, DC: The National Academies Press. Available from http://www.nap.edu/catalog.php?record_id=11658.

NRC. 2008a. *Emerging Cognitive Neuroscience and Related Technologies*. Washington, DC: The National Academies Press. Available from http://www.nap.edu/catalog.php?record_id= 12177.

NRC. 2008b. *Nanophotonics: Accessibility and Applicability*. Washington, DC: The National Academies Press. Available from http://www.nap.edu/catalog.php?record_id=11907.

NRC. 2009. *Persistent Forecasting of Disruptive Technologies*. Washington, DC: The National Academies Press. Available from http://www.nap.edu/catalog.php?record_id=12557.

2

Modeling, Simulation, Games, and Computing

INTRODUCTION

Ever-increasing computing power over the past 60 years has directly assisted the evolution of modeling, simulation, and games (MS&G), leading to new capabilities for developing models and simulating natural and engineered phenomena with greater realism and accuracy. In the meantime, the desire to model and simulate ever more complex phenomena has driven the demand for greater computing power, creating a feedback loop that has existed since the first computer.

Important shifts have been occurring within the relationship between computing and modeling, simulation, and games. Advances in semiconductor technologies are now delivering a teraflop (10^{12} floating-point operations per second, or FLOPS) of computing power on a single computer chip, a capability that once required significant infrastructure and was available only to nation-states. As teraflop capability on a chip becomes commercially available, nations and other actor sets will be able to simulate fairly complex phenomena with realistic three-dimensional geometry on desktop systems.

Moving to the next level of sophistication in modeling and simulation (M&S) will require integrating these teraflop chips into different architectures. As such the technological constraints and key technological trends that should be monitored are shifting to other domains—software and the supporting human capital.

This chapter discusses some of the key advances in computation and human capital lying at the intersection of M&S and games that will empower the MS&G of tomorrow. Additional discussion of these advances details potential risks to U.S. leadership and our ability to take advantage of them.

FIGURE 2-1 In 1945 the first general-purpose electronic computer was termed the ENIAC and worked principally to solve design problems with the hydrogen bomb. SOURCE: Image courtesy of Los Alamos National Laboratory/Science Photo Library.

THE PATH TO EXASCALE COMPUTING

The Evolution of Computing Architectures: From ENIAC to Multicore

There has been enormous growth in computing capability over the past 60 years, with an overall performance increase of 14 orders of magnitude.[1] Since the inception of the first general-purpose electronic computer, the Electronic Numerical Integrator and Computer (ENIAC), capable of tens of FLOPs (see Figure 2-1), the U.S. computer architecture research agenda has been driven by applications that are critical to national security and national scientific competitiveness. The most dramatic increase has occurred over the past 20 years with the advent of massively parallel computers and associated programming paradigms and algorithms.[2] Computing capability growth over the past 30 years and projections for the next 20 years are shown in Figure 2-2. Through the late 1970s and into the early 1990s, supercomputing was dominated by vector computers. In 1987 a seminal paper on the use of massively parallel computing marked an inflection point for supercomputing (Gustafson et al., 1988). Instead of the very expensive, special-purpose hardware found in vector platforms, commercial off-the-shelf parts could be connected with networks to create supercomputers (so-called Beowulf clusters).

Although the programming paradigm for these new parallel platforms presented a significant challenge, it also presented enormous potential. From the mid-1990s through today, massively parallel computers have ridden Moore's law (Mollick, 2006) to gain more performance for less capital cost.

Simulations using greater than 10,000 processors have become routine at national laboratories and supercomputer centers, while simulations using dozens and even hundreds of processors are now routine on university campuses. However, the computing future presents new challenges. The high-performance computing (HPC) community is now looking at several departure points from the past 15 years in order to leverage the increasingly common use of multicore central processing units (CPUs) and accelerators, as projected in Figure 2-2. Exascale initiatives are being developed by several federal agencies, and contracts for 10-plus petaflop computers have been awarded. Notable examples include the National Science Foundation's (NSF) Blue Waters system located at the National Center for Supercomputer Applications and the National Nuclear Security Administration's (NNSA) Sequoia system located at Lawrence Livermore National Laboratory. IBM is developing both systems within its Power 7 and Blue Gene product lines, respectively.

---

[1] Available from http://en.wikipedia.org/wiki/Supercomputing. Last accessed May 27, 2009.
[2] Available from http://en.wikipedia.org/wiki/Supercomputing. Last accessed May 27, 2009.

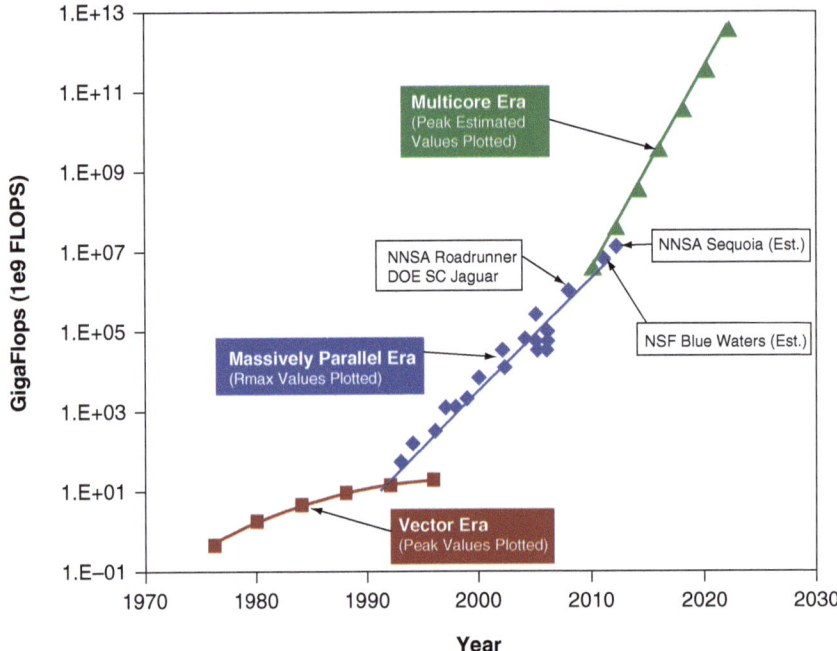

FIGURE 2-2 Rapid growth in computing capability spanning eras of vector, massively parallel, and future multicore trends. SOURCE: Vector era and massively parallel era data from Top 500 (http://www.top500.org/); multicore era data notional based on aggressive growth.

Multicore Processing

As component sizes continue to shrink while the processing speed of a CPU remains almost constant, vendors are adding additional processing units to a single chip, creating a new form of parallelism. Each of the processing units is termed a *core*; one example of a multicore chip is shown in Figure 2-3. Additional performance for a "processor" is being obtained by adding more cores, as shown in Figure 2-2. However, as demonstrated in Figure 2-4, clock frequency appears to be asymptotic, though the actual upper bound is still unknown. Multicore processors are currently achieving speeds higher than 3 Ghz.

Perhaps the most dramatic example of multicore chip technology was Intel's demonstration of a teraflop (10^{12} FLOPS) on a chip in 2007 (Figure 2-3). This remarkable achievement is clearly demonstrated when compared to the first teraflop computer, the NNSA's Advanced Simulation and Computing (ASC)[3] program Red Storm platform. The Intel processor is about the size of a dime and consumes 62 watts of power. A decade earlier, a peak teraflop of computing power required 2,500 square feet and consumed half a megawatt of power.

However, there is still much progress to be made to get the work performed on the Red Storm platform translated to the Intel chip or its counterparts, principally due to the limitation of the memory subsystem delivering data to the processor. While processors have doubled in performance approximately every 18 to 24 months in accordance with Moore's law, memory speeds have increased at most

---

[3]ASC is the abbreviation for the collaborative Advanced Simulation and Computing Program across Department of Energy (DOE) national laboratories to ensure the safety and reliability of the nation's nuclear weapons stockpile. Available at http://www.lanl.gov/asc/about_us.shtml. Last accessed June 15, 2009.

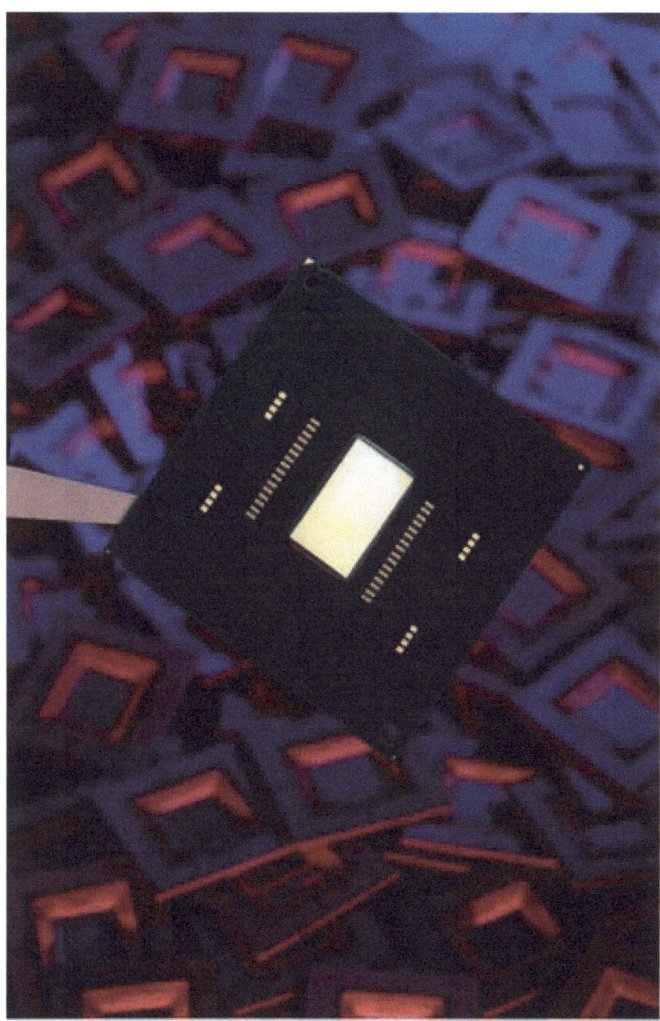

FIGURE 2-3 Intel's 1 teraflop on a chip (275 mm^2 and 62-watt power consumption), demonstrated in 2007. This single chip contains 80 cores, hence the term "multicore." SOURCE: See http://techresearch.intel.com/articles/Tera-Scale/1449.htm. Accessed May 27, 2009. Reprinted with permission of Intel Corporation.

10 percent per year. This has been termed the "memory wall" (Wulf and McKee, 1995). A fundamental reason for this slowed increase in memory speed is that, with present memory chip design, increased clock frequency results in increased power consumption (i.e., transistor leakage current increases with increasing clock frequency).

The mismatch between CPU and memory speed has meant that many applications, especially those in the national security arena, have been getting less and less of the peak performance from next-generation CPUs because the algorithms for many national security applications do not reuse the data in cache. Rather than ride Moore's law, these applications' performance increases are instead trending along the speed increases realized in memory systems. For example, in finite element simulations, very few bytes in a cache line are used once, much less multiple times, before writing back to memory. Overall, studies have shown that for many national security applications the processor spends the majority of its time doing integer arithmetic to determine memory locations (Moore, 2008).

Finding 2-1: "Cache-friendly" algorithms have been developed, but many take longer to run than "non-friendly" algorithms. For intelligence analysts responsible for technology warning, a major breakthrough in memory speeds would be a game changer and have significant national security implications. One

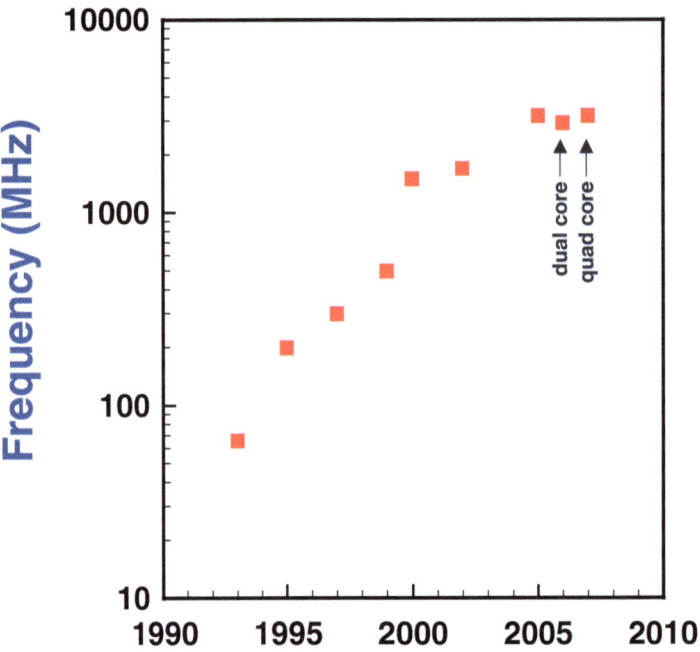

FIGURE 2-4 Clock speed in terms of frequency versus year. All data except those labeled refer to single-core processors. Even with additional cores added to a chip, the chip clock speed has currently reached an asymptote. SOURCE: Dunning (2009).

area showing promise is three-dimensional packaging of the memory and CPU, which provides for more pins, and thus higher bandwidth, to be connected between the two devices (Kogge et al., 2008). Rather than monitoring advances in processor speeds, tracking improvements in memory speed could provide earlier warning of the next step change in capabilities.

This technology assessment is summarized in Chart 2-1.

CPU Accelerators: Graphical Processing Units

Computer vendors are now exploring adding accelerators to CPUs, reminiscent of the vector era of computing in the 1980s and early 1990s (see Figure 2-2). The accelerator area has been dominated by the fast-paced gaming industry. Driven by the gaming community's appetite for faster and better three-dimensional graphics rendering, the computer industry has responded with unprecedented speed increases through graphics accelerators (NRC, 1997; Dunning, 2009).

Graphical processing units (GPUs, or graphics processors), a core technology for video-intensive games, are greatly outpacing commodity processors, as shown in Figure 2-5. Multicore GPU chips include the 240 cores on the $300 Nvidia graphics processor sold today with many desktops and laptops. The floating-point capability of Nvidia's, GPUs is almost four times that of the most powerful commodity CPU for the cheapest GPU, and newer GPUs such as Nvidia's Tesla S1070[4] are providing many

---

[4] Available at http://www.nvidia.com/object/product_tesla_s1070_us.html. Last accessed June 16, 2009.

CHART 2-1 Collapsing the Memory Wall

Technology	Observables
Memory speed: Rather than ride Moore's law for ever-increasing processor speeds, many applications' performance increases are instead following the speed increases realized in memory systems. A major breakthrough in memory speeds would be a game changer and have significant national security implications. Some applications could run 10 to 50 times faster. In the hands of an adversary, they could leapfrog the United States in simulation capability by 5 to 7 years.	A successful breakthrough in memory speed (and latency) will entail increased memory clock frequency without associated increased power consumption.

Accessibility	Maturity	Consequence
Level 2	Technology watch	Given notable increases in memory speeds, a greater fraction of theoretical peak performance can be achieved. With existing parts at existing speed but without the memory wall, a two- to threefold improvement would be possible.

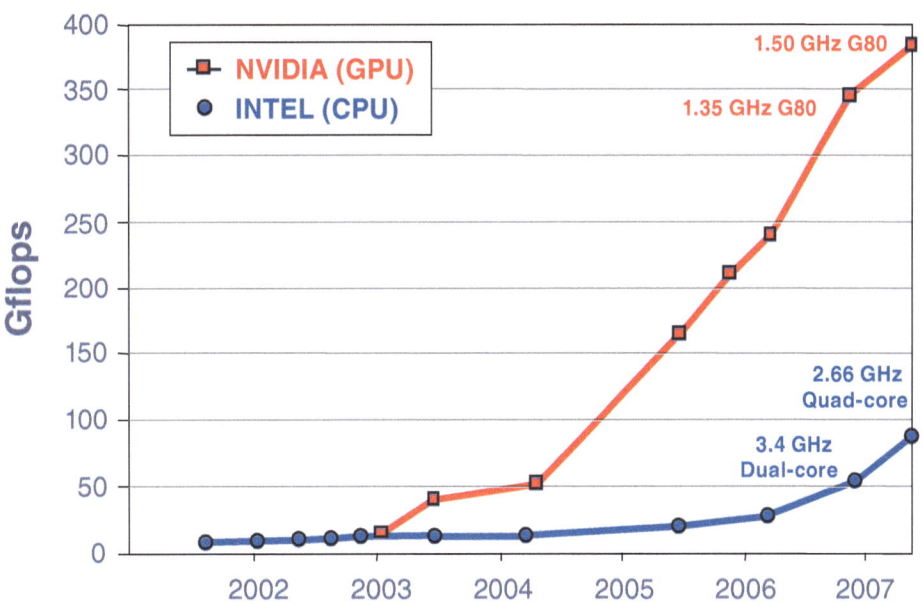

FIGURE 2-5 GPUs, a core technology for games, are greatly outpacing commodity CPUs. SOURCE: Dunning (2009).

FIGURE 2-6 NNSA's ASC Roadrunner supercomputer was the first general-purpose supercomputer to reach a sustained petaflop. SOURCE: LeRoy N. Sanchez, Records Management, Media Services and Operations, Los Alamos National Laboratory (http://www.lanl.gov/news/index.php/fuseaction/home.story/story_id/13602). Image courtesy of Los Alamos National Laboratory.

times greater performance. This trend is likely to continue, and commodity processor companies, such as Intel, have taken notice by investing in competing technologies such as Larrabee (Intel Corporation, 2008). As programming models and languages (e.g., Nvidia's CUDA[5]) evolve for GPUs, unprecedented performance gains will be realized for scientific software in the near future.

Already, hundreds of scientific software applications that are inherently data parallel and can take advantage of multiple cores have been ported to GPUs running CUDA, with speed-ups ranging from factors of 2 to a thousand.[6] Factors of 10 have huge impacts on scientific code, since larger systems and/or longer timescales can be accessed—often the difference between solving a problem and not solving it. Factors of 100 are game changing, enabling wholly new scientific questions to be addressed that were outside the typical capabilities available to many scientists. With factors of 100 to 1,000 in speed-up, state-of-the-art, three-dimensional scientific simulations that previously required an expensive supercomputer can be done on a desktop computer costing only a few thousand dollars or less. Financial analysis, medical imaging, computer vision, bioinformatics, materials design, real-time ray tracing, and virtual worlds will all greatly benefit from these trends.

Accelerators can, with significant effort, be programmed to solve scientific problems. Two examples of this can be found with the Sony PlayStation 3 (PS3). The Folding@home distributed computing program (Snow et al., 2002) uses Sony PlayStations throughout the world, contributed freely by their owners, to perform protein folding around the clock. This extremely latency-tolerant software exploits unused gaming cycles to advance biological science. Another example of PS3s advancing research is the ASC Roadrunner system, the first petaflop (10^{15} FLOPS) platform, shown in Figure 2-6. In this case, a tightly integrated, massively parallel platform was constructed with Advanced Micro Devices dual-core

---

[5]See http://www.nvidia.com/object/cuda_home.html.
[6]See http://www.nvidia.com/object/cuda_home.html#state=filterOpen;filter=Science.

Opterons, where each core has two IBM Cell processors (the same processor type as found in Sony's PS3) attached as accelerators. Although the Roadrunner system represents an impressive capability, the processor architecture represents a significant design departure from the past 20 years and will require significant rewriting of national security applications to take advantage of the accelerators.

CPU Accelerators: Field-Programmable Gate Arrays

Field-programmable gate array (FPGA) technology has found its way into the HPC market. The attraction of FPGAs is their electrical efficiency given their potential computing power. For example, the Xilinx V5LX330 is rated at 60 gigaflops and consumes only 15 W. As a comparison, the Intel Woodcrest processor (a two-core system) is rated at 48 gigaflops but consumes between 80 and 130 W.[7] The FPGAs obtain this performance through configurable logic—the same silicon can be reconfigured on the fly to support different operations and thus pipelining. Kernels of key national security algorithms have greater than 10 times and in some cases greater than 100 times increased performance. Harnessing these performance gains is limited due to an FPGA programming model that typically requires rewriting applications in hardware description language. Higher level language support is an active research area, and many universities and U.S. national laboratories have research programs to investigate the use of FPGAs for scientific applications, including creating HPC architectures that attach FPGAs as accelerators to the main processor.

Roadmap for Future High-Performance Computing

The high-performance (supercomputing) computer architecture research agenda in the United States, with stewardship falling primarily to the Department of Energy and the Department of Defense (notable programs include the NSF Blue Waters and the NNSA Sequoia systems previously described), will continue to be driven by applications that are critical to national security and national competitiveness. This leadership territory is no longer ours alone, and trends indicate that more countries will compete for the top 10 platforms in the future. Over the past 10 years, the United States has owned 50 percent or more of the top 500 supercomputers.[8] However, in November 2007, four systems in the top 10 were outside the United States, including a surprising entry from Indian commercial company Tata Sons. In November 2008, China joined the top 10 list with a system from Dawning. The take-away point from these two examples is that leading-edge HPC is now being pursued by the global commercial sector. The U.S. government and HPC industry have competition.

As described in a recent international assessment of simulation-based engineering and science by the World Technology Evaluation Center (WTEC, 2009), the world of computing is flattening, and any country with interest and minimal resources can have access to HPC. Several countries have the technology, resources, and desire to be first to achieve exaflop (10^{18} FLOPS) computing. Japan, Germany, France, and China are all committed to HPC, with ready access to world-class resources, faculty, and students. Germany is investing nearly U.S. $1 billion toward next-generation HPC hardware in partnership with the European Union (WTEC, 2009). Japan has always been a leader in HPC, and as the first to achieve 40 teraflops in the past decade with the Earth Simulator built by Japanese computer company NEC, Japan has an industry-university-government roadmap to reach exascale computing by 2025. In a recent turn of events, however, NEC and Hitachi, two of the three major vendors leading the develop-

---

[7] Available at http://rssi.ncsa.uiuc.edu/docs/industry/Xilinx_presentation.pdf. Last accessed May 27, 2009.
[8] Available at http://www.top500.org/. Last accessed June 22, 2009.

ment of Japan's Life Simulator (or *Kei Soku Keisanki*—a 10-petaflop computer planned for 2010-2011 to leapfrog past Blue Waters) pulled out of the partnership because of the 2009 global financial crisis which has caused huge losses and massive layoffs. This announcement leaves Japan's next-generation supercomputing project and its future HPC roadmap in question (Feldman, 2009). An NEC press release from May 2009 reads:

> NEC estimated that costs for moving forward with the manufacturing phase of the Next-Generation Supercomputer Project would significantly impact earnings for the fiscal year ending in March 2010, due to extensive investment required for the computer's manufacturing. Therefore, NEC has decided not to participate in the project's manufacturing phase. . . .
>
> NEC will now contribute to the project through the application of new vector architecture and optical interconnect technologies completed during the design phase, in addition to architectural research for rapid computing, development, education and operational support for application software with universities and research institutions. (NEC, 2009)

Without NEC and Hitachi as full partners, the future of Fujitsu's SPARC64 VIIIfx chip (the 128-gigaflop scalar processor it had been developing for the Life Simulator and HPC in general) also is in question. According to an HPCWire report (Feldman, 2009), the new Sparc chip, code-named Venus, is an eight-core CPU with an embedded memory controller and enhanced SIMD (single instruction, multiple data) support. The report describes claims by Fujitsu that the Venus prototype delivers more than double the raw floating-point performance of the Intel Nehalem processor while consuming much less power. However, without the partnership of NEC and Hitachi, support of the chip's development by the Japanese government is unclear. With no domestic computer vendors committed to petascale computing and beyond, the Japanese government and RIKEN (the government-funded research agency charged with leadership of the petascale computing project) will look to international partnerships to fulfill or redefine their roadmap.

Software applications will continue to evolve, in many cases rewritten to take advantage of petascale multithreaded platforms. Following the direction of Internet commerce, the M&S user space is expected to migrate toward virtual desktops. This is principally due to the reduced cost of managing application solutions. Application environments are anticipated to become like Google searches, with the computing infrastructure hidden from the user (e.g., cloud computing). Increased network connectivity, the pressure to lower costs, and highly replicated worker stations continue to drive globalization of software development and analysis teams. Such advances will eventually make access to petascale computing relatively ubiquitous, potentially enabling breakthroughs in a broad range of science and engineering disciplines, including molecular science, weather and climate forecasting, astronomy, earth sciences, and health (Ahern et al., 2007), and business and commerce.

The pace of the next major leap in computing power, an exaflop computer, will be significantly challenging to accomplish in a cost-effective and scalable way. The top obstacles are power consumption, resilience, the memory wall (Chart 2-1), and scalability. Kogge et al. (2008) discuss these areas in great detail. Using computer vendor roadmaps suggests that an exaflop computer would require finding 10^3 to 10^4 times more parallelism in existing applications, hundreds of megawatts of power, and a mean time to failure of only a few minutes. Power requirements and resiliency obstacles are a significant challenge and will need to be resolved by the HPC industry with government-supported research and development.

Finding the increased levels of parallelism for strong scaling will require new approaches to the underlying algorithms. *Strong scaling* holds the problem size fixed and improves computational speed by increasing the number of processors used to solve the problem, while *weak scaling* allows the problem size to increase by increasing the number of processors. In the era of teraflop to petaflop computing, weak scaling has dominated because of the requirement of realistic, numerically converged, three-dimensional simulations. In addition, for much of this era the processors were single-core chips and increased processing speeds provided much of the increased HPC capability. In exaflop computing, strong scaling, which improves computational speed, will be an important consideration for many time-dependent applications, as opposed to weak scaling that improves the ability to process larger problems.

Finding 2-2: Exaflop-level computing is expected to be aggressively pursued by the United States, Europe, and Asia (WTEC, 2009). Areas for intelligence analysts responsible for technology warning to watch that could facilitate a breakthrough for usable exaflop computing include technologies that:

- Significantly reduce memory power consumption;
- Dramatically increase memory-to-processor communication speed while reducing latency;
- Dramatically increase processor-to-processor communication speed while reducing latency;
- Dramatically reduce part counts through integration or packaging that will increase the mean time to failure; or
- Automatically generate scalable code.

Software for Exascale Computing

Expertise in writing scientific applications for massively parallel architectures is fairly limited and even more limited for accelerators like GPUs and FPGAs. Programming models, libraries, and compilers that automate parallelization and threading of sequential code would open these future HPC platforms to a broad set of application developers. Automated parallelization has been a holy grail of HPC since the appearance of autovectorizing compilers in the 1980s. For example, the Portland Group, Inc., is working on a compiler with open multiprocessing-like options that will autogenerate CUDA (Compute United Device Architecture) kernels from loops.[9] Although there are limitations, this work is a first step in the direction of compilers that may eventually enable nonexperts to obtain blazingly fast speeds on many applications.

High-level scripting languages (such as Python) that could be compiled to a GPU kernel would also make these accelerators more accessible to a wider range of users. Already, any programmer familiar with MATLAB, a popular computing environment used by university students around the world, can easily access the power of a GPU. Using a program called Jacket made by AccelerEyes, which runs under MATLAB (created by MathWorks), just a few changes to MATLAB m-files will cause the computationally expensive parts of the code to run on the GPU.[10] With well over 100 million CUDA-enabled Nvidia GPUs deployed worldwide in laptops and desktops,[11] such breakthroughs could have important national security implications (see Chart 2-2).

Algorithmic advances will be another key to sustained exascale computing. Improved algorithms have resulted in performance gains of several orders of magnitude in many application domains (Keyes et al., 2003, 2004), providing speed-ups that in some cases far surpass Moore's law with a doubling of performance every *eight*, rather than every 18, months (WTEC, 2009). An example of the role of algo-

---

[9] A loop in computer science is "a sequence of instructions that repeat either a specified number of times or until a particular condition is met," as defined at http://www.answers.com/loop. A description of this CUDA kernel generation process is available at http://www.pgroup.com/resources/accel.htm. Last accessed June 15, 2009.

[10] Available at http://www.mathworks.com/products/connections/product_detail/product_35939.html. Last accessed June 3, 2009.

[11] Personal communication between David Kirk and Committee Member Sharon Glotzer on July 16, 2009.

CHART 2-2 Software for Massively Parallel Architectures

Technology	Observables
Expertise in writing scientific applications for massively parallel architectures is currently fairly limited and even more limited for accelerators like GPUs and FPGAs. Experience programming explicit data movement technologies (e.g., vectors) is even more limited.	Monitor international HPC conferences and open-source software distribution sites for programming models, libraries, and compilers that automate parallelization and threading of sequential code.

Accessibility	Maturity	Consequence
Level 2	Technology watch	The ability to easily write applications for massively parallel architectures will open future HPC platforms to a broader set of application developers (e.g., Google, MATLAB).

rithms versus hardware in achieving major performance gains is highlighted in the ITER[12] Roadmap authored by Stephen Jardin (Princeton University) and David Keyes (Columbia University) (Sipics, 2006). According to a recent international study on simulation-based engineering and science (WTEC, 2009), to simulate the dynamics of ITER for a typical experiment would require computers with 10^{24} FLOPS. The Jardin-Keyes roadmap proposes how the 12 orders of magnitude required to bridge this gap can be achieved from today's teraflop computers:

> ... three orders of magnitude will be gained through hardware, with both processor speed and increased parallelism contributing equally; the other nine orders of magnitude will come from algorithms (software), namely four orders due to adaptive griding, three orders due to implicit time-stepping, one order due to high-order elements, and one order due to the reformulation of the governing equations in field-line coordinates. (Sipics, 2006)

The algorithmic advances attained in this fusion example will benefit the larger scientific community far beyond the fusion community. This example demonstrates the need for algorithmic advances in HPC in addition to hardware advances for modeling and simulation.

The key role that next-generation algorithms will play in achieving peta- and exascale computing implies that this domain will require close watch in order to determine potential breakthrough capabilities. The committee notes though that tracking advances in algorithm writing is significantly harder than tracking advances in hardware development.

Future Technologies Enabled by Exascale Computing

A study by Oak Ridge National Laboratories (Ahern et al., 2007) examined the potential scientific discoveries enabled by exaflop computing. A partial summary of that study's results is given in Table 2-1.

---

[12]ITER is an international research collaboration for fusion energy.

TABLE 2-1 Summary of a Study that Examines the Potential Scientific Discoveries Enabled by Exaflop Computing

Science Driver	Science Objective	Impact
Understand synthesis of alloy nanoparticles with potential impact for design of new catalysts	Define the thermodynamics of compositions of alloy nanoparticles	Magnetic data storage; economically viable ethanol production; energy storage via structural transitions in nanoparticles
Climate decadal climate prediction	Cloud-resolving (1-5 km) atmosphere; longer time integration (100-300 years, 1,000-year spin-ups); larger ensembles (5-20)	Understand and prepare for committed climate change
Dynamical linking of socioeconomic and climate responses	Couple infrastructure, climate, demographic, informational, and energy economic models to predict adaptation as communities react to stresses on infrastructure systems and propose potential policies	Identify future energy infrastructure needs
Systematic, large-scale exploration of optimal materials for catalysis or nuclear materials separation agents	Combine density functional theory with evolutionary search for complex materials or an accurate combinatorial approach to screen the best separation material out of O(103) compounds	Virtual design of catalysts and separating agents

SOURCE: Oak Ridge National Laboratory.

HUMAN CAPITAL IN SCIENCE-BASED MODELING AND SIMULATION

There is a profound lack of highly skilled computational scientists and engineers able to fully leverage the current HPC state of the art (Benioff and Lazowska, 2005; see also WTEC, 2009, and references therein), creating large-scale concern for a skilled scientific workforce capable of exploiting next-generation petaflop and exaflop systems. One major finding in a recent international assessment of simulation-based engineering and science (SBE&S) is as follows:

> Inadequate education and training of the next generation of computational scientists threatens global as well as U.S. growth of (the discipline of) SBE&S. This is particularly urgent for the United States; unless we prepare researchers to develop and use the next generation of algorithms and computer architectures, we will not be able to exploit their game-changing capabilities. (WTEC, 2009)

M&S is prevalent in every discipline and subdiscipline of science and engineering. Some fields, such as aerospace engineering and atmospheric science/climatology, have been totally transformed by computer simulation (Oden et al., 2006). At most major universities, students have access to a variety of courses in computational science and engineering, but many key topics related to "programming for performance" fall between the cracks, and much of the skill needed to program state-of-the-art supercomputers is still learned primarily through peer-to-peer learning and apprenticing. Strictly speaking, HPC is not computer science, yet it is also not a domain science. Thus from a curriculum standpoint, the core competencies of the multidisciplinary HPC field are often orphaned as neither computer science nor computer engineering, or other departments have direct responsibility for them. With new accelerator and multicore architectures, this lack of formalized curriculum is even more prevalent. Examples of core competencies for HPC include but are not limited to the following (Glotzer et al., 2009):

- Alternative architectures utilizing FPGAs and GPUs;
- Multicore/many-core/massive-multicore computer architectures for scientific computing;
- Modern languages and application programming interfaces for multicore architectures;
- Programming for performance (compilers, caches, networks, memory latency, throughput, and addressing, etc.);
- Debugging, performance, and validation tools;
- Big data input/output, storage, analysis, and visualization;
- Co-processor and acceleration utilization for scientific applications;
- Application development environments and application frameworks;
- Modern software engineering;
- Validation, verification, and uncertainty quantification; and
- Systems design, program management, testing, systems management, and systems maintenance.

In the late 1980s and early 1990s many universities created interdisciplinary certificate programs (and a few have since created Ph.D. programs) in computational science and engineering to address, in many cases successfully, the education and training gap that existed then. Many of those programs persist today. However, as the pace of computer architecture development has quickened, many of these programs have not evolved sufficiently, and faculty, students, and employers alike are frustrated by the limited M&S skills learned by scientists and engineers (WTEC, 2009). Furthermore, many universities lack a critical mass of expertise at the leading edge of HPC today. As a consequence, there is general consensus among faculty and employers around the world that today's computational science and engineering students are ill-prepared to create and innovate the next generation of codes and algorithms able to utilize new supercomputing architectures (including accelerators and multicore processors; WTEC, 2009).

The WTEC report also gives evidence that this lack of proper preparation begins in high school (or earlier) and continues through graduate education and beyond. Obviously in graduate school, increased topical specialization without concomitant depth in computing know-how further exacerbates this. In short, there is a clear and widening gap between what today's students learn vis-à-vis HPC and what they need to know to develop scientific codes for massively parallel computers and/or multicore architectures, including the graphics processors developed for the gaming industry and beginning to be used in science and engineering applications. Instead, it is generally thought that students use codes primarily as black boxes, with only a very small fraction of students learning proper algorithm and software development skills. Students receive no real training in software engineering for sustainable codes and little training in uncertainty quantification, validation and verification, or risk assessment, which is critical for decision making based on modeling and simulation.

These human capital issues may pose a potential threat to the United States by exploiting HPC for modeling and simulation in new and novel ways. Without addressing the threat, U.S. capabilities in terms of prediction and assessment will be limited both on the battlefield and off. As computers have become faster, models have increased in fidelity, causing games to become more realistic and accurate. Second, the increasing fidelity of games has led game makers to seek the manufacture of higher-performance computer chips. As such, the need for a skilled workforce in M&S, and in simulation-based engineering and science generally, will become increasingly important to national security to take advantage of improvements in speed and accuracy (Benioff and Lazowska, 2005; SIAM, 2001; SIAM, in press).

HUMAN CAPITAL IN COMPUTER AND VIDEO GAMES

While there is declining academic investment in high-performance computing, game design and development in the meantime together are becoming a new and accepted academic discipline worldwide, creating new interdisciplinary paradigms that will touch on computer science, simulation, and related disciplines. This is expected to create not only new consumer applications but also the "serious" gaming applications discussed in Chapter 3 (Iuppa and Borst, 2006).

Historically, the computer games industry began in the late 1970s as a hobbyist activity. Given the primitive state of microcomputers and graphics then, most games during the earliest days could be designed, developed, and produced by a single programmer working at home. These individuals were typically self-taught, working in a range of early microcomputer languages, including assembler, Pascal, and BASIC.

The growth of the gaming market in the 1990s included larger audiences demanding more sophisticated products, increased computational and graphics power in the home, and ultimately the rise of multiplayer games using large-area networks or the Internet. Collectively, these changes had several implications for the way games were created.

First, larger development teams had to be deployed. From the solo developer in the 1980s industry, the process grew to include literally hundreds of contributors for "A list" titles in the early 2000s. In a growing industry, this meant that both the budgets for individual game titles were growing exponentially and that talent had to be sourced and managed worldwide. In the 1980s, it cost tens of thousands of dollars to develop a game; in the 1990s, it cost millions; now it costs tens of millions.[13]

Second, these teams had to be organized among individuals with skill sets that previously would not have worked together. A partial list of professional categories typically included in a single project is given in Box 2-1.

Note that this list does not include the substantial studio or publisher overhead in the form of executive oversight, sales, marketing, public relations, and so forth. The cost to develop a top-line release from a major publisher increased from $50,000 in 1980 to $10 million or more in 2009 (Dobra, 2009). Not only has the risk become higher, but the need for talent, skilled management, reliable production processes, and controls, is also greater.

Ironically, the rise in mobile gaming and casual Web-based gaming has introduced a contrapuntal trend, as many mobile games and Flash-based browser games can be developed and launched online by one or two developers for a few tens of thousands of dollars. These games are typically free to play and supported, if at all, by advertising and micropayments from users.[14] Furthermore, casual gaming environments are increasingly mobile, socially networked, and enabled by products such as the iPhone/iTouch, which offers a large ubiquitous platform, leading to this form of gaming becoming increasingly viral in nature.

While the early games industry was able to build up a pool of accomplished developers and its early culture tended to foster informal transfer of knowledge to new entrants, the need for scalability in

---

[13]See, for example, http://www.411mania.com/games/news/121194/Cost-to-Develop-Gran-Turismo-5-Now-at-60-Million-Dollars.htm. Last accessed November 13, 2009.

[14]This "contrapuntive trend" is not unlike that between $200 million petascale computers and $200 GPUs, both of which will revolutionize science and engineering simulation.

> **BOX 2-1**
> **Partial List of Professional Categories Typically Included in a Single Game Development Project**
>
> Game designers
> Game producer (and assistant producers)
> Writers
> Director of engineering
> Technical directors
> Software engineers
> Graphics engineers
> Sound engineers
> Technical art director
> Lead art designer
> Three-dimensional artists
> Animators
> Two-dimensional artists
> Cut-scene artists
> Character artists
> Motion capture technical lead
> Motion capture artists
> Motion capture engineers
>
> Actors
> Voice-over talent
> Scripting (game-level scripting)
> Dialogue
> Video producer
> Video editors
> Sound engineers
> Musical director
> Musicians
> Tools development lead
> Tools developers
> Tools quality assurance
> Multiplayer engineering lead
> Multiplayer engineers
> Operations director
> Operations personnel (for online play)
> Quality assurance (typically 100 or more individuals worldwide
>
> SOURCE: *Call of Duty 4* credits found at http://www.mobygames.com/game/ps3/call-of-duty-4-modern-warfare/credits. Accessed June 19, 2009.

managing knowledge by the industry inevitably led to the formation of academic programs in games. These programs tend to fall into four categories:

- Computer science programs in game development;
- Humanities-oriented programs in game design and management;
- Media studies programs that research the impact of video games on individuals and society; and
- Trade school programs that prepare students for many of the other skills needed to enter the games industry (such as three-dimensional modeling, scripting, Web design, user interface design, sound effects, and music).

In 2000 the Independent Game Developers Association (IGDA) began a program to assess the state of emerging academic activity in game-related programs at universities worldwide. The purpose of this ongoing initiative "has been focused on setting curriculum guidelines, to enhance collaboration between industry and academia and to provide guidance to students wanting a career in games."[15]

In February 2008 the IGDA's Game Education Special Interest Group released a document entitled *IGDA Curriculum Framework: The Study of Games and Game Development* to assist academics and their institutions in establishing one or more undergraduate and graduate programs. The document provides

---
[15] Available from http://www.igda.org/academia/. Last accessed June 3, 2009.

TABLE 2-2 Number of Institutions Offering Postsecondary Courses and Degrees in Game Studies as of March 2009, by Country

Country	Number of Institutions	Country	Number of Institutions
United States	161	Colombia	2
Canada	49	Japan	2
United Kingdom	43	Norway	2
Australia	21	Switzerland	2
Sweden	10	Turkey	2
India	9	Argentina	1
Germany	6	Belgium	1
Brazil	5	Chile	1
France	5	Greece	1
Denmark	4	Hong Kong	1
Malaysia	4	Israel	1
Mexico	4	Italy	1
Singapore	4	Korea	1
Spain	4	New Zealand	1
Austria	3	Pakistan	1
Finland	3	Serbia and Montenegro	1
Ireland	3	South Africa	1
Netherlands	3	Taiwan	1
China	2	Thailand	1
TOTAL			367

SOURCE: Game Career Guide.

a generalized framework that outlines core topics (e.g., game programming, game design, business, critical game studies), degree programs, and institutional considerations for adoption of game-related courses and degrees.

As of March 2009 there were over 160 four-year undergraduate and graduate institutions in the United States that offer courses and/or degree programs in a range of game-related studies. Notable universities with comprehensive programs include the University of Southern California (with degree programs in both the computer science and cinema departments), Carnegie Mellon University, the Massachusetts Institute of Technology, the University of Michigan, Michigan State University, and the Rochester Institute of Technology.

International Presence

Overseas interest in academic programs for game development is increasing accordingly. According to lists maintained by the Game Career Guide,[16] there are well over 200 institutions in 37 countries offering a range of postsecondary courses and degrees in game studies. The breakout by country is listed in Table 2-2; breakout by region is listed in Table 2-3. These data evidence that game development is no longer just a phenomenon of the Western and English-speaking world.

English-speaking countries still dominate the academic games programs, with the United States, United Kingdom, Canada, and Australia dominating the field. Given the rise of the commercial game

---

[16]See http://www.gamecareerguide.com/schools/. Last accessed May 7, 2009.

TABLE 2-3 Number of Institutions Offering Post-Secondary Courses and Degrees in Game Studies as of March 2009, by Continent

Region	Total Schools
North America	210
Europe	92
Asia	48
Central and South America	13
Near East	3
Africa	1
TOTAL	367

SOURCE: Game Career Guide.

market in Japan and China, however—and noting its strong orientation toward online games (Mulvenon, 2008)—further growth can be expected in academic game programs in these countries.

Industry Implications of Formal Academic Game Programs

Given that there are several different areas of game curriculum (game design, game technology, trade school skills, game studies, and criticism), the effects of game curriculum programs on the games industry, and ultimately on the market and even national security, will be difficult to predict. Nevertheless, a few hypotheses seem plausible.

Games curriculum programs today are like computer science and engineering programs in the 1990s. There are now hundreds of such programs, both formal and informal, throughout the United States. Traditionally, development knowledge in the games industry was handed down from person to person in an informal mentor-apprentice fashion, usually as the result of working together on a project team. Formal computer science disciplines (such as artificial intelligence) were injected into specific games on an ad hoc basis. Games curriculums are likely to become more mainstream in universities as the specialty expands in understanding and professional membership. These programs are likely to improve overall rates of innovation as the membership and supporting community gain mass.

Because many game design/technology programs have a strong orientation toward project output (one University of Southern California graduate student's game *flOw* was the most downloaded game on Sony's PlayStation 3 network as of May 2008 (Sheffield, 2008), we are also likely to see a range of games produced that are not necessarily rooted in commercial entertainment needs. In particular, with academia's ability to attract government and foundation funding, there are already signs that college and university students are producing an increasing number of *serious games*—games that have explicit purposes such as learning, training, or deeper understanding of a specific subject through interactive play.

While formalized university training and popular games or hits are not necessarily correlated (or researched at this time), there is reason to believe that the university route to success in games is beginning to take hold. Valve's *Portal* game was sourced from a university project. Sony's *Cloud* game was originally a USC student project, and the development studio ThatGameCompany is staffed heavily with USC and Northwestern University graduates. As stated, there is no empirical evidence that such training is leading to a higher level of hits. Major publishers and developers are increasingly in need of highly talented developers and are sourcing them through universities as it is seen as fairly efficient even if additional training is needed. What may actually be driving university-based success though is not just the education but the fact that at such programs students have increased access to resources, time, peers, and mentorship devoid of commercial pressure to work on new ideas and forms of games. The better university programs also place strong emphasis on developing new theories and approaches to games that might not get the same cultivation within commercial game development. Thus, it is a combination of self-selection of talent into game programs and stronger education and mentorship, coupled with the environment that universities create, that is potentially turning them into potent incubators of talent and new games.

Almost all game development kits being produced now have application programming interfaces that allow access and the sharing of information via Facebook. For example, the iPhone SDK incorporates this feature, and many university students now build games connected to Facebook. This development will likely continue; an important issue will be whether Facebook has been designed with any scalability to take on this increasing burden. Second, further exploration of neuroscience research on the ways in which the brains of frequent gamers are noted to function differently with regard to perception, speed of response, and multitasking will be important. There are now companies focusing on low-cost hybrid EEG (electroencephalogram) sensors for interacting with and analyzing games and players (e.g., EmSense, NeuroSky, Emotiv). The biggest push for brain sensing is coming from advertisers who want to know what the consumer thinks of the way they advertise their brands.

FINAL THOUGHTS

Advances in computing technologies up to the teraflop level have yielded the raw computing power required to model and simulate fairly complex phenomena relatively ubiquitously. A threshold has been crossed where M&S capabilities that were once the province of nation-states and the more technologically sophisticated are now global and available to any and all actors. It is clear that no one society can maintain a sustainable, strategic, comparative advantage of access to these capabilities.

The next level of computing capability—petascale and exascale computing—will require new architectures with interesting implications for anyone concerned with the monitoring of technological progress. The use of massively parallel and multicore architectures raises algorithm and software development, rather than the development of hardware, as the key barriers to overcome. In fact, the demand for ever more powerful processors by the games and visual computing industry in order to deliver ever higher fidelity images is driving the evolution of much of the hardware (NRC, 1997). Less well understood is what advances in algorithms and software are needed to take advantage of massively parallel, multicore-based architectures, let alone who is training the human capital that will produce these advances.

From a technology watch perspective, this implies that, to progress to exascale computing and beyond, a shift in relative focus and resources may be in order, from tracking the progress in hardware development to that of algorithm/software development and human capital development. The fact that this latter goal is a fundamentally harder challenge is noteworthy.

REFERENCES

Published

Ahern, S., Sadaf Alam, Mark Fahey, Rebecca Hartman-Baker, Richard Barrett, Ricky Kendall, Douglas Kothe, O. E. Messer, Richard Mills, Ramanan Sankaran, Arnold Tharrington, James B. White, and the Computing Requirements Team of the National Center for Computational Sciences. 2007. *Scientific Application Requirements for Leadership Computing at the Exascale*. Oak Ridge, TN: Oak Ridge National Laboratory. Available at http://www.nccs.gov/wp-content/media/nccs_reports/Exascale_Reqms.pdf. Accessed May 27, 2009.

Benioff, Marc R., and Edward D. Lazowska (co-chairs, President's Information Technology Advisory Committee). 2005. *Computational Science: Ensuring America's Competitiveness*. Available at http://www.nitrd.gov/pitac/reports/20050609_computational/computational.pdf. Accessed June 19, 2009.

Dobra, Andrei. 2009. Ubisoft: Game development will cost $60 million in the future. *Softpedia*. Available at http://news.softpedia.com/news/Ubisoft-Game-Development-Will-Cost-60-Million-In-the-Future-114487.shtml. Accessed June 19, 2009.

Feldman, Michael. 2009. NEC, Hitachi bail on 10-petaflop supercomputing project. *HPCWire*. Available at http://www.hpcwire.com/blogs/NEC-Hitachi-Bail-on-10-Petaflop-Supercomputing-Project-45047917.html. Accessed June 17, 2009.

Glotzer, Sharon C., R. Panoff, and S. Lathrop. 2009. Challenges and opportunities in preparing students for petascale computational science and engineering. *Computing in Science and Engineering* 11(5):22-27.

Gustafson, John L., Gary R. Montry, and Robert E. Benner. 1988. Development of parallel methods for a 1024-processor hypercube. *SIAM Journal on Scientific and Statistical Computing* 9(4):609-638.

IGDA (International Game Developers Association). 2008. *IGDA Curriculum Framework: The Study of Games and Game Development*, V2.8 beta, Technical Report. Mt. Royal, NJ: IGDA. Available at http://www.igda.org/wiki/images/e/ee/Igda2008cf.pdf. Accessed May 7, 2009.

Intel Corporation. 2008. First Details on a Future Intel Design Codenamed "Larrabee." Press release, August 4. Available at http://www.intel.com/pressroom/archive/releases/20080804fact.htm?cid=cim:ggl|larrabee_us_brand_t|kEB2B|s. Accessed April 5, 2009.

Iuppa, Nicholas, and Terry Borst. 2006. *Story and Simulations for Serious Games: Tales from the Trenches*. St. Louis, MO: Focal Press.

Keyes, D., P. Colella, T. H. Dunning, Jr., and W. D. Gropp, eds. 2003. *A Science-Based Case for Large-Scale Simulation*, vol. 1. Washington, DC: U.S. Department of Energy. Available at http://www.pnl.gov/scales/. Accessed June 17, 2009.

Keyes, D., P. Colella, T. H. Dunning, Jr., and W.D. Gropp, eds. 2004. *A Science-Based Case for Large-Scale Simulation*, vol. 2. Washington, DC: U.S. Department of Energy. Available at http://www.pnl.gov/scales/. Accessed June 17, 2009.

Kogge, Peter M., Keren Bergman, Shekhar Borkar, Dan Campbell, William Carlson, William Dally, Monty Denneau, Paul Franzon, William Harrod, Kerry Hill, Jon Hiller, Sherman Karp, Stephen Keckler, Dean Klein, Robert Lucas, Mark Richards, Al Scarpelli, Steven Scott, Allan Snavely, Thomas Sterling, R. Stanley Williams, and Katherine Yelick. 2008. *Exascale Computing Study: Technology Challenges in Achieving Exascale Systems*. Atlanta: Georgia Institute of Technology. Available at http://users.ece.gatech.edu/~mrichard/ExascaleComputingStudyReports/ECS_reports.htm. Accessed June 22, 2009.

Mollick, E. 2006. Establishing Moore's law. *IEEE Annals of the History of Computing* 28(3):62-75.

Moore, Samuel. 2008. Multicore is bad news for supercomputers. *IEEE Spectrum*, Nov. Available at http://www.spectrum.ieee.org/computing/hardware/multicore-is-bad-news-for-supercomputers. Accessed July 14, 2009.

Mulvenon, James. 2008. "True is false, false is true, virtual is reality, reality is virtual": Technology and simulation in the Chinese military training revolution. *The "People" in the PLA: Recruitment, Training, and Education in China's Military*, Roy Kamphausen, Andrew Scobell, and Travis Tanner, Eds. Strategic Studies Institute of the U.S. Army War College; Carlisle, PA, pp. 49-90.

NEC. 2009. NEC Reviews Participation in the MEXT Next-Generation Supercomputer Project. Press release, May 14. Available at http://www.nec.co.jp/press/en/0905/1401.html. Accessed June 17, 2009.

NRC (National Research Council). 1997. *Modeling and Simulation: Linking Entertainment and Defense*. Washington, DC: National Academy Press.

Oden, J. Tinsley, Ted Belytschko, Jacob Fish, Thomas J. R. Hughes, Chris Johnson, David Keyes, Alan Laub, Linda Petzold, David Srolovitz, and Sidney Yip. 2006. *Revolutionizing Engineering Science Through Simulation: A Report of the National Science Foundation Blue Ribbon Panel on Simulation-Based Engineering Science*. Arlington, VA: National Science Foundation. Available at http://www.nsf.gov/pubs/reports/sbes_final_report.pdf. Accessed July 1, 2009.

Sheffield, Brandon. 2008. Finding a new way: Jenova Chen and ThatGameCompany. *Gamasutra.com*, May 5. Available at http://www.gamasutra.com/view/feature/3648/finding_a_new_way_jenova_chen_and_.php. Accessed June 25, 2009.

SIAM Working Group on CSE in Education. 2001. Graduate education in computational science and engineering. *SIAM Review* 43(1):163-177.

SIAM Working Group on CSE in Undergraduate Education. Undergraduate computational science and engineering education. *SIAM Review*, in press.

Sipics, Michelle. 2006. Taking on the ITER challenge: Scientists look to innovative algorithms, petascale computers. *SIAM News* 39(7). Available at http://www.siam.org/pdf/news/1005.pdf. Accessed June 22, 2009.

Snow, C., H. Nguyen, V. S. Pande, and M. Gruebele. 2002. Absolute comparison of simulated and experimental protein-folding dynamics. *Nature* 420(6911):102-106.

WTEC (World Technology Evaluation Center). 2009. Sharon C. Glotzer, Sangtae Kim, Peter T. Cummings, Abhijit Deshmukh, Martin Head-Gordon, George Karniadakis, Linda Petzold, Celeste Sagui, and Masanobu Shinozuka, panel members. *International Assessment of Simulation-Based Engineering and Science*. Baltimore, MD: WTEC. Available at http://www.wtec.org/sbes. Accessed June 22, 2009.

Wulf, William A., and Sally A. McKee. 1995. Hitting the memory wall: Implications of the obvious. *Computer Architecture News* 23(1):20-24.

Unpublished

Dunning, Thom. 2009. "Future of High Performance Computing." Presentation to the committee on January 12, 2009.

3

Games: Beyond Entertainment

INTRODUCTION

Digital games have a larger societal role than simply as pieces of stand-alone software entertainment. Like television in the 1950s and 1960s, games are part of mainstream culture and help describe and define "tribal" relationships within that culture.[1] Whether committed or casual gamers, players of all stripes find themselves shaped by the games they play and the ways they play them. Games hold the potential to broadly impact society as a learning tool, a social mechanism, a cultural translator (facilitating communication between individuals of different cultural backgrounds when a common language or cultural context is lacking), a productivity tool, and even a means to communicate messages and encourage change. Emerging developments such as integrating real-world information into game media demonstrate how games may revolutionize means for communication and analysis in areas as varied as social interaction and national security.

As mentioned in Chapters 1 and 2, there has been recent dramatic progress in high-performance computing (HPC) with clear relevance to modeling and simulation. While petascale and exascale architectures are unlikely to affect games directly, further advances in multicore and memory speeds are certain to affect future development cycles of the games industry.

As the intelligence community and others think about positioning themselves to utilize and understand games, they must first understand the range of game types and experiences and also the varied and complex ways in which players interact with games. Beyond recognizing this range of available gameplay "tools," described in depth in Appendix E, communities need to understand and take advantage of the wide variety of game platforms and potential applications beyond the entertainment world.

---

[1] For example, modern tribal relationships can be seen in players of games such as *World of Warcraft*, similar to television tribal relationships of the past such as those among Star Trek Trekkies.

GAMES AND CULTURE

Interactive and Participatory Culture

Games are heralded for being interactive media, even more so as video games have become more sophisticated not only in the area of technology—for example, artificial intelligence, physics, and graphics—but also in the area of game design. Digital games have evolved as exemplars of participatory culture where users not only play the games but also often contribute to further iterative developments. This participatory game culture yields a variety of added resources, in the form of new content in the game for others to play (e.g., new levels, new objects, new characters, environments) and outside materials that support the play of the game (e.g., walkthroughs and value databases).

It is important to recognize that in some ways games provide an innovative dimension of "co-creation" between the audience (gamer) and the storyteller (game designer). Beyond download sites where users can actively upload, download, and browse each other's content, propagation and dissemination of community-generated content has been automated in games like Electronic Arts' Spore, in which each end-user's single-player game world can be autopopulated with creatures that other players have created in their own games. In this fashion the game developer is able to increase the number and diversity of in-game assets at little or no marginal cost (because it is the players, not the game company's artists and animators, who create the assets). This trend is likely to become increasingly significant for the broad base of game development, in both entertainment and serious games, as the technologies and tools continue to evolve and more powerful computational platforms become available at lower cost.

It is worth noting that such innovations in game design are not dependent on either HPC or, indeed, any increase in core computing power. They are innovations in the social dimension of games that leverage existing, non-cutting-edge capacities of the Internet, and the commodification of storage and bandwidth. As game platform life cycles become longer and gaming moves increasingly online, the trajectory of the industry is no longer a technological arms race, as it was in the 1980s, 1990s, and early 2000s. Some of the most compelling and engaging experiences are and will continue to be delivered on relatively low-cost machines (e.g., the Nintendo Wii), making their influence widely accessible.

Social Interaction Through Games

For many years the popular perception of digital games has been that they are solitary experiences enjoyed by those who shun social interaction, such as introverted teenage males, in so-called real life (a term common in the domain of virtual reality). These are mistaken assumptions, as digital games have evolved to be highly social environments that appeal to a much broader demographic.

Because of both their persistent nature and sheer scale, massively multiplayer online games (MMO games, or MMOGs) support a level of social complexity and social interaction that is surprisingly rich and diverse.[2] One product of this is the presence of groups that self-organize both for the sake of socialization and for game play, collaboration, and competition. In most cases, MMOGs support these groups (which are commonly referred to as "guilds" for MMOGs or "clans" for smaller-sized competitive online shooters) with in-game features (such as guild management and guild chat) designed specifically to facilitate the organizational behavior. The most dedicated guilds often have a complex social structure with leaders (who determine policy and planning), officers (who implement policy), and regular members. Further, the work of scholars such as Williams (2006; Williams et al., 2008) and Yee

---

[2] See Appendix E for a more complete description and discussion of genres of MMOGs, such as MMO role-playing games like *World of Warcraft* and MMO simulation games like *Second Life*.

(2006, 2009) suggests that the bonds formed between players in an MMOG—individuals who rarely, if ever, meet each other face to face—are held in as high a regard as and are as lasting as bonds formed between friends in real life. Likewise, the structure of complex guilds often collapses due to the same pressures that cause traditional social groups to collapse (Ducheneaut et al., 2007).

Once gamers identify the categories of games in which they are most comfortable, they begin to learn the value system and allowable attitudes of their particular game genre's culture. Norms and values are strictly enforced within the guilds and often vary from guild to guild. While there is certainly room for cheating and other aberrant behavior in all games, there is usually a built-in and externally applied retribution.

While MMOGs are often cited as examples of complex social interaction in video games, the social interaction associated with single-player games should not be dismissed. Many online game services that act as a platform for the digital distribution of single-player games (the most notable being the Valve's *Steam* game service) feature robust social networking tools that allow players to build friend lists, chat with other players, and rate games. These tools create communities of social interaction around the single-player game experience.

Game-based online connections for both single-player and MMO games have outlets to become offline connections, as demonstrated in some cases by the prevalence of conventions. Such events as PAX, GenCon, BlizzCon, and Comic-Con International act as venues for both the construction of fan-based video game culture and the social interaction of players themselves. The participatory culture of games has grown as games have evolved to be both multiplayer and performance experiences. In multiplayer games, gamers socialize in groups online for both collaborative and confrontational experiences. Research has shown that, while many online gamers will play with seemingly random players, a large portion actually play with people who are in close geographic proximity with each other[3] and, in the case of many, with people they know from real-life interactions (Williams et al., 2008). Many people derive stature within both game communities and other social communities by virtue of praise gained from interest in their game-related contributions (Chen, 2009).

Player Engagement

The existence of games that benefit from the play of other humans is in itself a form of participatory culture. In many cases, the player's role is not simply presence but something specific such as group leader, socializer, or other player who provides some level of performance in a game that creates its own level of entertainment or organization (Tychsen and Hitchens, 2009); further discussion of player roles can be found in Appendix E.

Performance experiences are relatively new in gaming but are becoming a major contributor to the socialization and participatory culture of games. Performance games are those titles where being seen as "good" at the game includes not only how well you play but also how superlative your performance is while playing. A good nongame analogy might be the judgment of presentation for a gourmet chef in addition to the pure taste of the dish. Performance games are often gestural in nature (e.g., *Guitar Hero, Rock Band, Wii Tennis, DanceDanceRevolution*) where the real-world movements of the player become part of the game itself. While the games require real-world performance to play, the actual level of performance required for game success is a subset of what players often contribute. In essence, players "overperform" to the game's minimum requirements for proper input, but in overperforming they

---

[3]The committee notes that some of this regional localization is forced by server protocols and balancing and latency of chip multiprocessors.

gain socially in the eyes of their peers. The value of performance game play cannot be judged outside a socialized setting, as those values are granted outside the game by fellow players, further reinforcing the social experience of these games.

Online Community Contributions

Many games are affiliated with planned online communities. Online communities serve to provide additional intellectual property value to the game as a place where game fans can both purchase game-related gear and find game updates and post requests for game fixes and extensions. These extension suggestions to games are often quite valuable to companies. On the *America's Army* game, there were some 200,000 subscribers to the online forums in its first year, many of whom sent in desired game fixes and new level suggestions (Carter, 2002). Sometimes companies even provide mod (modification) tools so that fans can produce additional levels for their favorite game. These mods produce strong relationships between the game developers and the player community, the fans essentially having "skin in the game" with the developers, as the developers benefit from the creativity (and cheap labor) of the players from expert to novice. Additionally, such modding capabilities may hold unrealized potential for nongaming communities.

Transmedia and Popular Culture[4]

Games are a largely user-controlled combination of software development (i.e., procedural algorithms that include modeling and simulation) and media (e.g., writing, film, art, music, storytelling). While this may be true of other sectors (such as simulations and some multimedia), games are often far more advanced in their usage of both software and media to produce an engaging experience. Games' status as a bridge to other media (e.g., film, music, comic books) allows interactive media to draw on these other forms, aesthetically and culturally, in an interactive environment driven by increasingly sophisticated technology from graphics processors to artificial intelligence.

Gamers move across multiple media forms as part of their utility of play. Players of *Pokemon* will play the offline paper card game, read user-created *Pokemon* how-to guides, or watch the *Pokemon* cartoon all in an effort to obtain information they can use to improve their performance in the game itself. *World of Warcraft* players will routinely turn to a variety of out-of-game media, including map books, walkthroughs, object databases, message boards, and more, in an effort to improve their game knowledge and thus perform better.

In many of today's popular entertainment franchises (especially speculative fiction and fantasy), stories and characters unfold across multiple media channels and products. As Jenkins (2008) notes, the strength of transmedia storytelling lies in the fact that multiple texts are integrated into an overall narrative so large that it cannot be contained within a single medium. In its ideal form a transmedia story speaks to the strengths of each individual media platform. A story might be introduced in a film and expanded through television, novels, and comics, and then its world might be explored and experienced through digital game play. Each franchise entry needs to be self-contained enough to enable autonomous consumption. That is, you do not need to have seen the film to enjoy the game and vice versa. However,

---

[4]Transmedia is a phenomenon that has existed for many years. Japan was one of the first countries where toy companies, comic book publishers, video game manufacturers, and media companies created formal partnerships and property holding companies for new products. Some examples include Teenage Mutant Ninja Turtles, Pokemon, Bakugan, and Mighty Morphin Power Rangers. In the United States, media companies such as 42 Entertainment have emerged as specialists in transmedia entertainment and marketing.

there also needs to be a strong narrative connection between the separate manifestations of the story in the separate media platforms. Whether they start as a video game and migrate to other media (e.g., the films *Final Fantasy*, *Tomb Raider*, *Silent Hill*) or start as noninteractive media properties that are made into games (e.g., *Harry Potter*, *Batman*, *SpongBob SquarePants*), popular characters and fantasy worlds bridge multiple media and must be understood as a cross-media phenomenon.

Ultimately, the best kind of transmedia storytelling is not about telling the same story over and over again but about telling different parts of the same story in different media in order to create a meta-story that transcends any one specific delivery platform. Perhaps the best example of this from digital games is *Enter the Matrix*. Developed by Shiny Entertainment and published in 2003, *Enter the Matrix* complemented the North American theatrical releases of *Matrix Reloaded*, the final movie in the *Matrix* trilogy. What is important about *Enter the Matrix* was that it was never intended to be a simple game adaptation of the film, allowing players to experience the same story and the same action that took place in the movie. Instead, it was intended to be part of a larger story ecosystem that included not only the films but also several short anime movies released on DVD. Each piece in this story ecosystem fits together in order to create a much larger story experience. The game itself allowed players to play characters and events that were only briefly mentioned in the films but played an important role in the holistic understanding of the ongoing meta-story.

In fact, this transmedia phenomenon has reached such a level that many creative storytellers such as Peter Jackson (director of *Lord of the Rings* and *King Kong*) are beginning to think about multiple media channels not so much as an opportunity to repeatedly tell the same story or one large story but instead as an opportunity to create holistic storytelling realities in which many different stories can be told. In the most successful cases, what wraps each of the stories in a transmedia franchise together is a storytelling reality that is specifically designed (and evolved) with continuity and canon in mind. The reality connects the stories and creates a layered storytelling matrix of interlocking events, people, economy, philosophy, technology, and culture that is not only fed by the stories but in turn feeds the stories, making them part of a living world. While fictional realities such as J. R. R. Tolkien's Middle Earth fantasy, Frank Herbert's Dune reality, and H. P. Lovecraft's Cthulhu mythos were once rare exceptions, many creative storytellers are imagining and building worlds that serve as expansive storytelling sandboxes.

In short, games are a mainstream element of mass market popular culture. The realization of these transmedia connections as a business trend is described further in the section "Games and Transmedia as Marketing" later in this chapter.

Political and Other Simulation Games

The cultural role of games is not limited to socialization and cross-media realizations; they can also work to simulate human networks and even yield persuasive power on the game player. A government or political simulation is a game that attempts to simulate the government and politics of all or part of a nation.[5] The related genre of "god games" combines the systematic thinking of real-time strategy games with an underlying simulation of real-world tasks. In these games the player takes a god-like perspective and possesses supernatural powers to intervene (constructively or destructively) in the development of an interconnected ecosystem or narrative environment whose characters are invariably small and (compared to the player) powerless creatures with limited finite state machine autonomy. Examples include Will Wright's *Spore* and *The Sims* and Peter Molyneux's *Populous* and *Black and White*.

Slightly distinct from the god games genre are simulation (or "sim") games, including *SimCity* and

---

[5]Available at http://en.wikipedia.org/wiki/Government_simulation_game. Last accessed June 29, 2009.

its sequel titles *SimEarth* and *SimAnt,* and business or political simulations like *Railroad Tycoon* and *Balance of Power*. In simulation games, as in god games, the player assumes a supervisory role over a virtual world that incorporates models of real-world complex phenomena (e.g., urban planning, biological and geological systems, economies, or geopolitical conflicts). Simulation has more leeway in terms of adherence to subject matter in the games world than in traditional modeling and simulation fields. To some extent, use of the terms "simulator" and "simulation" is often associated with those games with higher levels of task verisimilitude, while the prefix "sim-" is often used in games that tend to be more abstract ideas of simulation.

In some cases, the underlying models of simulation games incorporate assumptions that may be intentionally or unintentionally persuasive to the player. For instance, in *SimCity* it is impossible to create a thriving city above a certain size unless one incorporates and fosters public transportation into the urban infrastructure. This, in effect, makes a politically fraught argument for the benefits of public transportation. A simulation game based on models of, say, the federal budget or the effects of lifestyle and preventative care on one's physical health, such as *BudgetHero* (2008) or *SimHealth* (1994), would inevitably reflect the assumptions and biases of whoever designed the game. By playing to win the game, a player would at least temporarily buy into the designer's assumptions and role play based on those assumptions. Such examples demonstrate the persuasive potential of the genre. For certain military and vehicle simulation games, such as *Flanker, Falcon, Gran Turismo,* and *Forza,* a premium is placed on hyperrealism rather than driving a particular viewpoint.

OUTPUTS AND EFFECTS OF GAME PLAY

Context of Game Play

Recent attention in the games research community has been paid to situational context: the idea that games as media and as software are exceedingly adaptable and plastic. Given this flexibility, games take on very different properties depending on the context in which they are used and applied, from changing surroundings to play motivation to audience. Game play experience changes dramatically when a game is played in a mobile setting on a bus versus on a big screen at home driven by a powerful game console; with direction in a classroom compared to an educational game played alone without a mentor available; or for the chemistry training of highly motivated individuals compared to that of unmotivated high school students.

Such situational context potential is often missed by those examining games for the first time. As the uses of games multiply, recognition of the context of a given game is essential. This includes understanding the reaction and positioning of a game across different cultural contexts. For example, in the United States, the action game *America's Army* is generally considered a realistic, doctrine-based look at the life of infantry in the U.S. Army and despite recruitment purposes is viewed in a mostly positive light as a "simulation" game (Kennedy, 2002). In some cultures, however, it is simply considered propaganda, and in others it is seen as a direct assault on their beliefs (Game Politics, 2007). This phenomenon is not confined to *America's Army.* Many Western-developed first-person shooters are positioned as Western (primarily U.S.) forces in a struggle against terrorists, often of Middle Eastern or Islamic descent. The combination of common Western themes in commercial games and the rapid uptake of games for military and security training in the West is beginning to elicit specific responses from organizations in other parts of the world seeking to push back or compete with such activity (Axe, 2008).

Situational play context can be defined beyond game-play context and cultural interpretations. Games are dependent on their enabling playback devices and distribution and marketing support. As such systems become more ubiquitous, video games will exist in many different situational contexts and locations. How these surrounding aspects can create change and enhance or depress potential impact, of either a particular game or games in general, is a critical issue that needs more examination, understanding, and development. As social networking platforms expand in their reach and other user-contributing "Web 2.0" systems grow up alongside games, the combination of games with these systems that create new situational constructs needs to be examined interdependently. Further examination of individual games in isolation cannot provide the full story of their impact; however, further exploration into the ways in which games take advantage of new mobile platforms, such as the iPhone, which combines media, computing and graphics power, kinetics, geolocation, networking, voice, and video, would potentially be rewarding.

Understanding the Effects of Game Play

For those who research, play, and advocate about the potential of video games, games are understood to generate outputs beyond the eradication of lazy Sunday afternoons. There is evidence from the existing research base so far to say that games produce outputs other than entertainment, such as improved hand-eye coordination and memory (Haier et al., 2009; Mayo, 2007).

Researchers continue to seek to quantify these effects by answering questions like:

- How much impact do games have in various areas of effect?
- What are the best development methodologies for instigating various outputs?
- What are the potential comparative advantages of gaming versus other approaches for effect? How can return on investment be compared between these approaches?

While some may consider game playing to be simply frivolous, research counters this view. As studies begin to unlock the effect games can have on individuals and groups of players as well as the means by which to instigate such outputs, the question becomes less "Do games create outputs beyond fun?" and more "How can we harness these outputs to the highest level of desired effect?" for whichever sponsor seeks such results. Certainly such motivations have driven the advertising industry, for example, to increase in-game ad placement for their products.

As game players and the population at large better understand what games offer in terms of byproduct from play, games will begin to shape a new set of products aimed at those changes people may seek. This is already happening in the health and wellness segment where *Wii Fit*—a game ostensibly for exercise—has sold close to 20 million units worldwide (Mazel, 2009), making it arguably equivalent to one of the largest "new drug" launches in the health space. PopCap games, a maker of small-sized casual games, explains that a large number of its players self-report using their games to relieve stress, improve memory, and avoid cravings for food or nicotine (Ng, 2009). This suggests the commercial market may impact the techniques developers hone to author games generating extra-entertainment outputs.

A possible classification for the types of nonentertainment outputs and effects from games might break down as follows:

- Learning and skills,
- Biological and physiological health,
- Psychological and behavioral health,
- Communication and organizational behavior,
- Productivity, and
- Advocacy and propaganda.

Learning and Skills

Currently the largest amount of work goes into the learning and skills area as it relates to the creation of games to support knowledge acquisition, the acquisition and practice of new skills, and literacy and understanding (Griffiths, 2002; Yates, 2008).

For many the quest to understand alternative uses of games is about seeking new tools for knowledge and skill acquisition (Durlach et al., 2000; Goerger et al., 1998; Okagaki and Frensch, 1994). Many sponsors of games aim to provide specific populations with the ability to acquire and hone skills through game play. Advocates of games, when asked to highlight the types of skills and talents players can gain through game play, often highlight general life skills such as creativity, planning, strategy, and collaboration. However, these skills are inherently difficult to assess. The lack of acceptable testing for creativity, for example, makes it difficult to scientifically verify whether these skills are further developed in people by games than they would be otherwise.

At the same time, for skills where reliable evaluation metrics exist, the exact impact of games and simulation is still an open question. For example, mathematical skills—that is, skill in manipulating mathematical expressions in contrast to skills associated with new mathematical insights—may rise through playing a math game, but the significance of these gains may be debatable, especially when compared to other learning methodologies. In such specific targets the issue of underlying motivation arises. For many learners, game playing is a pleasurable way to study. Motivation is a critical component of learning, but it is not enough. In fact, some math games developed so far have at times been demotivating due primarily to poor design and software development and insufficient budgets (Elliott et al., 2002; Rice, 2007).

Biological and Physiological Health

In the health sector, research shows how game play can be responsible for direct changes to human biology and physiology. Games have been shown to potentially increase visual processing skills,[6] spatial awareness (Subrahmanyam and Greenfield, 1996; Feng et al., 2007), balance (Brumels et al., 2008), endothelial function (Murphy et al., 2009), and—through "exergaming"—aerobic utilization and weight loss (Murphy et al., 2006). Games have also been developed to allow patients to visualize attacking cancer cells.[7]

Psychological and Behavioral Health

Games have been used to combat attention deficit (and hyperactivity) disorder, to develop dexterity in limbs and fingers (*Medical News Today*, 2008), and to improve reaction times in the elderly (Dustman et al., 1992). As the committee heard in the course of preparing for this report, one group is even looking at how specialized games can be developed to aid in the neurological development of young children by directly honing specific skills acquired in their formative years in parallel to specific biological developments of their central nervous systems (McBride, 2009). Applying games as a means of improving various aspects of human performance could become a more widespread, determined practice.

---

[6] A list of relevant publications related to the impact of gaming on visual development is available at http://www.bcs.rochester.edu/people/daphne/publications.html. Accessed June 4, 2009.

[7] See *Ben's Game* on the Make-a-Wish Foundation Web site, http://www.makewish.org/site/pp.asp?c=bdJLITMAE&b=81924. Last accessed October 15, 2009.

Communication and Organizational Behavior

Most of the outputs discussed thus far have focused on individual change. Games can provide group changes as well. As gamers play together in groups large and small, they gain specific new ideas in how to communicate, organize, and act in adhoc collaborative environments, a skill that will be in increasing demand in a global transient workplace.

Productivity

One output of game play that is only recently gaining public attention is the idea of productivity (BBC News, 2003; McDonald et al., 2007). This is the notion that using the motivational and interface capabilities of video games, tasks, and procedures can be codified and translated into programs that people will interact with in game-like fashion. As a result, the goal is an increase in the productivity of the underlying tasks or a worker's general daily output.

To the extent such work can be harnessed, games may also create entire new forms of work, management, and collaboration. Current examples in this area have included protein folding via the *Foldit* game at the University of Washington,[8] image labeling through the Amazon *Mechanical Turk* program, and software debugging (Pontin, 2007).

Comparing Gamers and Nongamers

If games are creating different outputs besides just entertainment, then to some extent we might begin to see a difference between avid players of video games and those who do not play them (e.g., "Nintendo surgeons," described in Satava et al., 1995). While the work has a long way to go, some studies have suggested notable differences between the two groups. For example, Dmitri Williams of the University of Southern California has published data showing that players of some massively multiplayer online games have a higher correlation to lower body mass indices (BMI) than the population at large (Williams et al., 2008). Researchers at the University of Rochester in New York have documented differences in visual processing skills between players and nonplayers of video games (as listed in footnote 6). Educators such as James Paul Gee of Arizona State University, Kurt Squire of the University of Wisconsin, and others have repeatedly brought up the differences between how problem solving is treated in games compared to in other more rigid and less forgiving cultures and organizations (Gee, 2003; Squire, 2003).

Despite these recognized differences, the direction of causality is not clear. In all these cases, there were significant differences between players and nonplayers but with only theoretical explanations. For example, Williams hypothesizes that the reason players as a population have a lower BMI is that they are exposed to less unhealthy advertising (Williams et al., 2008). Others argue the cause of lower BMI is because hands are more occupied during game play than passive television viewing, leading to reduced intake of snacks (Giammattei et al., 2003). It is also possible that gamers come from higher socioeconomic backgrounds and thus have more healthy food and exercise choices available to them. Clearly there have been many differences between populations of gamers and nongamers, and some may be directly related to game play. For now, however, researchers are still identifying such differences, and much more work is needed to see if the cause itself is video game play.

Overall, understanding video games and their outputs is in a state of emergence. The research described here indicates that video games can play a role in affecting human physiology, biology, thought, and capabilities. Gaming advocates argue that it is possible to hone these capabilities and drive them

---

[8]Available at http://fold.it/portal/. Last accessed June 29, 2009.

through gaming to a broad population. Organizations considering games as a tool must pursue specific research goals in order to reach definitive conclusions about the effects of game characteristics on players. Resources can then be properly allocated to produce a game that leverages the desired positive effects on targeted and larger populations.

Democratization

At their most basic level, games are combinations of software and media developed for consumer-level hardware systems. Games may be considered to offer a form of democratization as they make accessible powerful technologies, interfaces, content, simulation, and possibilities to wider populations of people and reduce hierarchical barriers to access. For example, Microsoft's *Flight Simulator* is an extremely capable flight simulation system built for low-cost desktop computers. Twenty years ago it would have cost millions of dollars to get a comparable system; in contrast, today, *Flight Simulator* is essentially a $30 investment on top of perhaps a $1,000 computer investment. Despite a constant (high) cost for the developer across both scenarios, in the commercial games market that cost is distributed across a broader consumer base. Today's PlayStation 3 (PS3) and Xbox could power their own flight simulators with great fidelity for under $500.

The concept of democratization means that games may hold opportunities for transcendent capability. Instead of providing small groups of people with basic instruction, it is possible to empower millions to train at a high level of instruction through game-based simulations run on low-end computer and mobile infrastructure. This is essentially the mission some nongovernmental organizations have begun to pursue for games, though without serious funding thus far.

Currently, the primary indicator of democratization has been the general spread of electronic entertainment itself. In examining this expansion, though, there are signs that video gaming is reaching populist levels of accessibility. In Japan the sheer volume of people playing brain fitness titles (despite various degrees of demonstrated efficacy) is in the millions (Kibb, 2007). Around the world, popular games spread very quickly and spur the creation of well-organized fan communities that create many augmented pieces of information, software, and other content in support of their favorite titles. These communities tend to be self-run even when supported by the software's developers, offering powerful examples of how gaming communities can rapidly self-organize. Such communal ecosystems are great potential resources, but developers and researchers are still learning to master the delicate balance between self-organized and self-policed communities and those who wish to exploit these communities' existence for their own purposes. A good example of this is *Falcon 4*, where user groups self-organized to modify the simulation and fix all of the bugs after the game was no longer distributed or supported by the publisher. From an early version of the source code that was leaked, the community created its own version of management and configuration control. Interestingly, the original design team detuned some of the flight and weapons capabilities from the game based on concerns that the product had the potential to depict highly sensitive military capabilities. Ironically, the modders fixed many of these weapons and flight capabilities. A group went so far as to organize all of the modifications, licensed the rights from the publisher, and formed its own company. The company commercially republished the game 7 years later under the name *Falcon 4: Allied Force* (see Wikipedia references to Falcon 4.0 and Falcon 4: Allied Force). The rereleased version not only had improved graphics and fixed bugs, but the modders also significantly improved the weapons, flight, and artificial intelligence capabilities. The rereleased version was a commercial success and was accomplished without any support from the original development team.

SERIOUS GAMES

Given the variety of possible outcomes from games, including entertainment, there are subsequently many reasons people play games. The purpose of play can be defined broadly as the interactions created implicitly by the developer/sponsor and those interactions created and communicated explicitly. In the case of most games, the implicit and explicit purpose of the game is entertainment, leisure, pleasure, and/or fun (these words are often used interchangeably by many developers and players but are often interpreted more granularly by researchers). With some games, however, there may be different intents between the explicit purpose of a game and its implicit purposes. The widening uses of games can be grouped into three main categories:

- *Player-directed:* Games are being used by players for purposes besides fun, entertainment, and leisure.
- *Third-party-directed:* Games are being repurposed by third parties (and not the developer or original sponsor) for their agenda (e.g., teachers reusing games like *Civilization* to teach history, *Madden* to coach football, and *Flight Simulation* to teach pilots).
- *Developer- and sponsor-directed:* Games are being created by developers and sponsors from scratch to achieve specific nonentertainment purposes.

Including and beyond these three categories, nonentertainment purposes span a much bigger range of situations and desired outputs than commonly assumed (see Table 3-1). Games with these expanded applications are referred to as *serious games*, a term that arose from books like *Serious Games* (Abt, 1970) and from the Serious Games Initiative of the Woodrow Wilson International Center for Scholars

TABLE 3-1 Serious Usage Segments

		Gamers: Serious Stuff Gamers Do	Third Party: Instructor, Therapist, Mentor, Leader, or Marketer	Developers: Commercial, Indie, Serious, and Software and Hardware	Technologies: Middleware and Tool Vendors (hardware and software)
Serious commercial, off-the-shelf (COTS)	Reset purpose	Wii exergaming	Curriculum development	Teaching guides and support forums	Machinima
	Modify software or hardware	*CounterStrike*	Revolution (Neverwinter Nights)	Nonentertainment modes of play	PS3 Folding@home
	Augment	*Guitar Hero* Sequencer maps for *WoW*	Curriculum development	PowerGrid SmartBrainGames	Linux on PS3
Serious games	Virgin development	XNA Creators Club, Independent Artist Development Group	Use of specialized serious games	Serious game COTS and specialized serious game development	Microsoft ESP or Breakaway MOSBE

SOURCE: Data provided by Ben Sawyer, Digital Mill, Inc., and Peter Smith, Joint ADL Co-Lab.

(started in 2002). Despite recent recognition of serious gaming, the practice goes back decades in the computer game field, and certainly the lineage of such activity includes their nongame counterparts that go back centuries. What is happening now, however, is a decided acceleration of this activity fueled by such factors as the Internet, e-learning advancement, social networks, computer graphics and processing capabilities, increased computational power, and the global communications network. This is also furthered by the overall growth in video games to a major global industry that channels billions of dollars into the development of new games, related software, services, and hardware.

Sponsors and developers of many serious games craft the explicit purpose of the game to be a specific desired output (e.g., learning or exercise) while also claiming to retain entertainment, leisure, pleasure, and/or fun as implicit purposes of the game (important for retaining or obtaining the player's engagement toward achieving the explicit purpose and thus outcomes of the game). This split role has also been reversed at times. Will Wright, creator of *SimCity* (1989), *The Sims* (2000), and recently *Spore* (2008), has often said that there are many implicit motivators for learning inside his games built for entertainment (Wright, 2006). Many indie and hobbyist game creators use games to embed commentary on current events or other targets (such embedded commentary was previously discussed under "Political and Other Simulation Games," and further examples are given in "Political Manipulation Through Games on the Internet" in Chapter 4). When third parties adopt essentially commercial entertainment games for new purposes, they redefine the purpose of the original game through the application of additional actions, techniques, or materials or by changing the context of game play.

Cultural Attitudes Toward Serious Games

For both game players and creators, the majority of attention to video games is oriented toward entertainment. Video game companies, many with a history of providing nonentertainment products, have often positioned themselves away from association with nonentertainment applications. There has been concern, especially in the North American market, that such association could damage their brand with core gamers or that their company could be the target of a shareholder lawsuit.[9] Cultural attitudes toward the purpose of gaming in Japan have been different. Because personal computers are not as prevalent in Japan as they are here and in Europe, many game consoles have featured educational games and content. In Europe also there are more games and game-like media oriented toward nonentertainment areas and simulation (Falstein, 2009). The United States is now seeing a shift back to a wider array of games as offered through the Wii console and other systems. Hit-product producers are realizing that this greater range of products, services, and offerings is not damaging to their brands and is in fact reaching largely untapped markets and communities.

What this arguably means is that a wholehearted embrace and effort toward extracting output from games beyond entertainment could provide some level of strategic initiative toward a goal such as education or widespread skill training by an institution or a state that promotes such activity. Given the speed at which software moves around the world, it is difficult to say how long such an initiative would last, what its economic gains would be, or what parts might sustain. While there has been research and calls for action to pursue the fullest potential of games for education, health, and otherwise (Thai et al., 2009), no single region or country currently prioritizes extracting maximum capabilities from the gaming medium for their broader benefit. Most activity thus far has either been via nongovernmental organizations (e.g., the Robert Wood Johnson Foundation and the MacArthur Foundation) or state-sponsored

---

[9]Private communication from an Electronic Arts vice president to Michael Zyda, May 2005.

funds—especially in Korea (Cowan, 2009) and France[10]—that are more orientated toward economic development outcomes than educational, productivity, training, or health outcomes.

Educational and Training Dimensions of Serious Games

Many universities are now building game development programs inside or aligned with their computer science departments (Barnes et al.), 2008; Horswill and Novak, 2006; Parberry et al., 2005; Phelps et al., 2006; Sung, 2009; Whitehead, 2008; Wolz and Pulimood, 2007; Zyda et al., 2008). There is now even a conference—the Foundations of Digital Games. While these programs are focused on teaching game development, they are also all initiating research efforts on games (Zyda et al., 2009). Research in games aligned with game development programs is important in that it allows us to experiment with games such that we can understand how to properly deploy this new interactive communications medium. Major research directions in games include work in infrastructure, cognition, and immersion (Zyda, 2005). Infrastructure research on the underlying software and hardware necessary for developing interactive games includes massively multiplayer online game architectures, game engines and tools, streaming media, and mobile platforms like the iPhone.

Cognition is research on the modeling and simulation of computer characters, story, and human emotion. Included in this is real-time analysis of game play, innovation in game genre and play styles, and integration of pedagogy with story. Immersion is research on creating technologies that engage the player's mind via sensory stimulation and providing methods for increasing the sense of presence. Included in this is advanced work on computer graphics, sounds, haptics, affective computing, sensing human state and emotion, and advanced user interfaces.

The Department of Defense has been moving toward using games since publication of the National Research Council report *Modeling and Simulation: Linking Entertainment and Defense* (NRC, 1997). That important study indicated to DoD that it needed to adapt games as the deployment medium of next-generation modeling and simulation systems, and DoD followed that recommendation. While DoD has deployed games for training, modeling, simulation, and analysis, it has yet to fund basic research to grow games beyond what they are today. Most funding for games goes to contractors whose sole goal is to build a precise game as requested by the DoD funding sponsor. In fact, some of the largest DoD funding in games has gone to non-U.S. contractors (Buxbaum, 2009).

Finding 3-1: As game development further becomes its own formal discipline taught in universities and is merged with existing modeling and simulation programs, the result will be a generation of practitioners more apt to apply innovations from game design and development to the larger modeling and simulation community.

THE BUSINESS OF GAMES

Digital games as a business have attained fast-growing success since their initial appearance in the 1980s. The dramatic growth of the games business and diverse facets of its business models emphasize the pervasiveness and still unrealized potential of the games industry. Further details on the industry's structure and evolving business models are discussed in Appendix E.

---

[10]Information on the serious games projects of the Chamber of Commerce and Industry of the Valenciennois (CCIV) is available at http://www.valenciennes.cci.fr/frame.htm. Last accessed July 9, 2009.

TABLE 3-2 Evolution of the Modern Games Industry

Era	Time Frame	Comments
Proto-console and PC games	1980s	The advent of computers jump-started the modern games industry. The 1980s also saw much experimentation with dedicated console devices.
Nintendo revives market	Late 1980s	Learning from the mistakes of Atari and Mattel, Nintendo launches the Nintendo Entertainment System and revives the console market by enforcing strict quality control and improved business models.
Consoles take over	1990s	Upon Nintendo's success, Sega, NEC, Sony, a new version of Atari, SNK, and others battle for console dominance. These wars eventually give way to just Sony and Nintendo, which are joined by late-entrant Microsoft. All consoles become a huge segment of the games industry.
Connected consoles	2000s	More powerful, networked consoles come to market and challenge the boundaries of what is possible (Xbox360, Playstation 2/3, Nintendo Wii).
Online and mobile games	2000s	The emergence of downloadable or Web-based games comes into its own as one of the largest forms of consumer entertainment.
Mobile ascent	2000s	Apple releases the iPhone and—combined with renewed Internet capabilities for Nintendo's DS handheld and a new wave of smart phones powered not only by Apple but also by Google's Android platform—the mobile gaming world takes off.

Evolution of the Games Industry

The modern games industry is roughly 30 years old and has come a long way in its short existence. The industry has gone through a number of important platform changes, economic cycles, and technology shifts on its way to becoming a major source of consumer entertainment. Table 3-2 summarizes some of the key evolutionary steps in the development of the modern games industry.

The economic impact of the modern global games industry continues to grow by leaps and bounds, as illustrated in Figure 3-1. In 2008, according to Edge: The Global Game Industry Network, the global games industry generated nearly $54 billion in revenue, up from $25.4 billion in 2004.[11] Estimates call for the global games industry to grow to nearly $68 billion by 2012, an expected growth of 9.1 percent annually from $41.7 billion in 2007.

Of $41.7 billion in 2007 video games revenue, $17 billion[12] was generated in the United States, making this country one of the largest individual markets for video games. On a regional basis, the Asia Pacific region is the single largest regional market for the games industry, accounting for some $14 billion in 2006 industry revenues.[13]

---

[11] Available at http://vgsales.wikia.com/wiki/Video_game_industry. Last accessed January 7, 2010.
[12] Available at http://xbox360.ign.com/articles/846/846392p1.html. Last accessed January 7, 2010.
[13] Available at http://www.ida.gov.sg/News%20and%20Events/20050704144250.aspx?getPagetype=20. Last accessed January 7, 2010.

FIGURE 3-1 Worldwide video games industry revenues from 2004 to 2008. SOURCE: Data derived from http://vgsales.wikia.com.

Global Industry Trends

The committee has observed a number of growing global trends in games, as described below. Such trends not only have immediate impact on the pervasiveness and cultural impact of games, but in many cases may open the door for new technologies and applications beyond games.

Rise of MMO Games

Massively multiplayer online games continue to gain momentum as a force in the market. No single game in the modern era has been more successful than *World of Warcraft*[14] (*WoW*; Blizzard Entertainment, Inc., 2008). *WoW* (see screenshot in Figure 3-2) is an immersive MMOG in which players band together to complete quests, battle other players, and navigate a complex and ever-changing world. With over 12 million active players and over $1 billion in revenue in 2008 (Securities and Exchange Commission, 2009), *WoW*'s commercial market is on a par with Activision's total revenue from its console games division.

MMO games and their corresponding "persistent worlds" have changed the landscape of gaming in several key ways. Because they are networked and online, they allow players from around the world to play with and compete against each other. *WoW* also supports multimodal communications, including voice-over-Internet protocol and instant messaging within the game. *WoW* players organize themselves into guilds that can persist over time and bridge the online and offline worlds. While *WoW* attracts players from around the globe, it is especially popular in China (Alexander, 2009).

The trend of MMO games has several degrees of relevance for national security. On the one hand, games like *World of Warcraft* can be seen as unmonitored global communication networks that require further analysis. Additionally, massively multiplayer online role-playing games in the context of military war games may provide insights into new effective campaigns for terror, psychological operations, and

---

[14]*World of Warcraft*® is a registered trademark and copyrighted product of Blizzard Entertainment, Inc., and is used with permission.

FIGURE 3-2 Screenshot of *World of Warcraft*. SOURCE: Courtesy of Blizzard Entertainment, Inc. World of Warcraft, Warcraft and Blizzard Entertainment are trademarks or registered trademarks of Blizzard Entertainment, Inc. in the United States and/or other countries. Image used by permission. All rights reserved.

mass reaction to certain catalysts that can be developed in online environments to provide rapid evaluation of new tactics, techniques, and procedures. Such opportunities are discussed further in Chapter 4.

Games and Transmedia Marketing

Discussion earlier in the chapter showed that games often exist as part of a cross- or transmedia entity. Just as games can work to extend plotlines from books and movies, games are used to market and expand access to other media. These games might be specially developed promotional alternate-reality games meant to build awareness for movies (Stephen Spielberg's *A.I.: Artificial Intelligence*), albums (Nine Inch Nails' "Year Zero"), or even other games (e.g., the western-themed digital game *Gun* was promoted by an online card game tied to the Old West's saloon culture). Some games serve as a promotional venue for other media, such as the songs used in soundtracks of sports or racing games. Games themselves can be an advertising medium, whether as promotional "advergames" on Web sites or as publishing properties (i.e., certain racing and sports games where marketers' logos are placed in the stadium or alongside the track). Conversely, Coca Cola riffed on *Grand Theft Auto*—setting a commercial inside a three-dimensional computer-animated world that resembled the popular video game—in one of its splashier 2006 commercials. Volvo actually used a popular racing game's graphics to cross-promote its cars and the game in a single broadcast ad.

Transmedia storytelling also makes sense from a broader economic standpoint. Different media attract different audiences. Films, television, and literary fiction probably have the most diverse audiences, while comics and video games have a less diverse audience (though that is certainly beginning to change). A well-conceived and well-crafted transmedia franchise attracts a wider audience by pitching the content differently in the different media (Jenkins, 2006). If each story product offers a fresh experience to the player/reader/user/audience, a crossover market will expand the potential gross within any individual medium. So people who may not play video games but enjoyed the *Lord of the Rings* movies, for example, might experiment with a related game title. Ultimately, any given product is an entry into the franchise as a whole, providing new levels of insight into the meta-story.

Intersection of Social Networks and Games

As more and more games add online capabilities, players are demanding greater ability to connect and communicate with their fellow players for coordination and socialization. Modern games are beginning to embrace the elements of social networking that have been successful, and games themselves have become some of the most popular applications on social networks. According to developer Analytics,[15] as of June 2009, seven of the top 10 daily use applications on Facebook were games. As Facebook, MySpace, Xiaonei, 51.com, and other leading social networks continue to grow in size and scope, these two markets will continue to collide.[16]

In the course of this integration, gamers find themselves with an increasing number of tools and technologies that allow them to connect with their real-world friends within games and to branch out and make new friends among game players. The ability for gamers to create new, persistent bonds with their fellow gamers makes games more social and a fertile place for users looking to make and deepen friendships. Additionally, businesses may be formed and run by people who have never met in person, who may have met initially through games or other social network applications, and who co-create complex products and collaborate fully from distributed locations.

Mobile Games Platforms: The Nintendo DS and the iPhone

Mobile games have been around since the start of the cell phone era (approximately 1994) but have largely been overshadowed by competing platforms. However, there have been a number of breakthroughs in the past few years that have reinvigorated the mobile games market in several demographics.

First, the Nintendo DS (dual screen), shown in Figure 3-3, sold over 100 million units globally in its first four years as a commercial product (Martin, 2009) and has put an innovative, powerful, low-cost game platform into the hands of many people. Nintendo DS developers have found a robust market for games that take advantage of the DS's capabilities for networking, touchscreen, and innovative game play.

The second major development, affecting a much different demographic, has been the launch of the Apple iPhone. The iPhone has proven to be one of the fastest-growing platforms in the mobile space, with 40 million iPod Touch and iPhones in circulation worldwide as of the Apple Worldwide Develop-

---

[15]Available at http://www.developeranalytics.com/search_app.php. Accessed June 29, 2009.

[16]For additional information on the volume of users of some popular social networking Web sites, including Facebook, MySpace, and Twitter, see http://www.facebook.com/press/info.php?statistics; http://siteanalytics.compete.com/myspace.com/; and http://www.businessinsider.com/chart-of-the-day-twitter-worldwide-uniques-2009-9. For a monthly comparison of users of MySpace and Facebook, approximately 180 million users overall, see http://siteanalytics.compete.com/myspace.com+facebook.com/. Accessed September 28, 2009.

FIGURE 3-3 A Nintendo DS handheld console. SOURCE: Courtesy of Wikimedia Commons. Image available from http://commons.wikimedia.org/wiki/File:Nintendo_DS_Lite_side.jpg. Accessed October 21, 2009.

ers Conference in June 2009 (Apple, Inc., 2009). Even more telling, of the roughly 50,000 applications available for download in the Application (or App) Store, roughly 40 percent are games.

The iPhone has very advanced technical capabilities—highly accurate global positioning system and location positioning capabilities; 3G (shorthand for the third generation of telephony standards, with the fastest performance available at the time of this publication; and Wi-Fi access, a touchscreen, an accelerometer, and emerging connections to social networks. The iPhone has all of the ingredients required to become a leading next-generation games platform, and developers are just beginning to explore how to best take advantage of the platform's capabilities. The recently released 3.0 SDK (software development kit) for this device has a large number of new application programming interfaces to support games and connections to social media like Facebook and YouTube. The Apple business model for the iPhone—allowing owners to easily download a wide variety of independently generated applications, including games—has been so successful that it is being copied on other personal electronic devices such as Blackberry and Android-based smart phones. Additionally, a major driver of the iPhone's success as a gaming platform is its economic model, where close to 70 percent goes to the game developer and 30 percent to Apple,[17] and the disambiguation of the traditional publishers, distributors, and retailers. The capital, publication approvals, and marketing barriers to publishing are very low on the iPhone compared to other platforms like the PlayStation Portable and the Dual Screen. This, in turn, has driven a resurgence of mobile games.

Mobile games on platforms like the iPhone and its successors put pervasive connectivity and interactivity into our pockets and with us continually. Such systems may allow a person on the ground to be part of the decision processes explored in online universes. (See Appendix E for a discussion of the mobile games platform for further analysis.)

The Microtransaction-Based Business Model

The latest business model innovation to hit the games industry is the advent of so-called free-to-play games. The free-to-play games industry allows players to enjoy some portion of a game for free without requiring a subscription or packaged software purchase. Free-to-play games have gained the greatest traction in places such as Korea (Ashby, 2009), where games cannot be supported by advertising, game console penetration is low, and broadband connectivity is plentiful. By removing the requirement to pay before playing, free-to-play gaming has allowed the games industry to broaden its market.

The free-to-play business model is still in development. At present, most free-to-play games generate revenue by allowing users to purchase in-game items (power-ups, decorative items, etc.) for small amounts of real-world currency (known in the industry as *microtransactions*). In the aggregate, many small purchases can add up to a very substantial market. Some of the top free-to-play games in Korea, such as Nexon's *KartRider* and *MapleStory* franchises, generate $100 million to $200 million in annual revenues (Wagner, 2009). In the United States, Zynga is the fastest-growing microtransaction online gaming company, with $50 million in revenue in 2008 and $100 million expected in 2009 (Chowdhry, 2009) via social media games on platforms such as Facebook.

The microtransaction phenomenon may become relevant outside the games realm as well. To get around credit card issues for many of these transactions (especially when young people and other populations without credit are factored in), many of these providers earn funds through the sale of prepaid cash cards for their games and by "point" purchases using mobile payment solutions.[18] The growing prevalence of such transactions may change the means by which a particular user can be tracked via transactions associated with a specific game account, with ramifications in defense and law enforcement.

International Competition in Game Development

The games business has always had a significant amount of international competition and this is currently on the rise. Japan has arguably been the leader in the home console space since the mid-1980s although it is important to note that the core central processing unit and graphics architecture of all three major systems come from U.S.-based companies (IBM, NVIDIA, AMD) and that the number two

---

[17]Information on this economic model can be found on the Apple Web site, http://www.apple.com/iphone/developing-apps-video/. Last accessed November 10, 2009.

[18]Prepaid cards and "points" also get around some highly litigious patent cabals, driving companies like Microsoft to use them instead of credit cards and direct withdrawals from bank accounts.

console is now from Microsoft. In the handheld space the console handheld space has been dominated by Japanese companies (Sony and Nintendo), but Apple's iPod line has become a major entrant in the space and is disrupting what had been a once-stable business for Nintendo especially. The iPhone is also upending what has been a major market for mobile phone handsets previously dominated by a combination of European-led companies (Nokia, Erricson) and Korean manufacturers (Samsung, LG) as well as U.S.-based Motorola.

In general terms, North America is the largest gaming market in the world, with Europe coming in second and Japan third, but is critically potent in successfully exporting its work to the other markets. Recently, the Japanese games market has slumped, forcing Japanese game makers to look increasingly to the West to raise profits. This has fueled, among other things, SquareEnix's purchase of U.K.-based SCI/Eidos over U.S. interests such as Warner Brothers. In some respects, despite its dominance in the console space, Japan has increasingly found itself on the defensive recently.

As the games industry matures and evolves, its general three-market dominance is also shifting. Korea, which has significantly invested in its broadband infrastructure, is becoming a bigger and bigger player in games, especially in Asia. Its developers have focused largely on multiplayer and massive multiplayer gaming on personal computers (consoles were not allowed in Korea until recently because of anti-Japanese import laws and cultural animosity over World War II) and an innovative business model where many core parts of games are given away for free while players who wish to upgrade make many microtransaction purchases for in-game items such as avatar fashions and special power-up items. This business model, now commonly called "Free 2 Play w/Microtransactions," has lent itself extremely well to lower-income markets in not only Asia but also the United States, where the "tweenage" and teenager market in some respects is a lower-income market.

Like Korea, China has also seen a robust PC-based games industry thrive. As migration over the past 20 years swamped cities, PC-based gamerooms became the most prevalent form of entertainment because they were easy to set up. Unlike movie theaters, sports stadiums, or amusement parks. PC game rooms and Internet cafes could be set up with no special permits or land requirements. This has given gaming a major infrastructure advantage over other forms of entertainment that Westerners commonly take for granted. Chinese gamers are similar to Korean gamers in that they have flocked to major MMOG and multiplayer games on PCs. China's games development scene has not produced the same level of major export force as Korea (which has brought forward major publishers like NC Soft and Nexon) but is quickly rising. Already Chinese studios owned by Japanese and Western interests produce a number of major console titles. China is also home to several partner distributions in the MMOG market, one being The9 Limited, which will likely begin developing and releasing content to the United States and elsewhere eventually. Instant messaging and gaming leader Tencent has an installed base of users that rivals Facebook.

Beyond Korea and China, Asia also has some rising development tide in Singapore (which has an experimental development lab in conjunction with the Massachusetts Institute of Technology), Taiwan, and Vietnam. On a more global level, Brazil is starting to become a major market, and is beginning to slowly see established studios providing work. Russia and Eastern Europe have always been a stalwart development market, even if a little uneven in their success rate.

India has been one of the more vexing markets for games thus far. Consoles are available and are beginning to sell but remain too expensive for many. Development studios are beginning to see some outsourcing capability, especially for art assets, but most developers and publishers still see more potential than the level of realization they have achieved in China. This may be due to the fact that India's information technology activity has taken away a lot of programming talent from areas that might be useful for games and also India's entertainment infrastructure has been more available to others versus what happened in China during the past 20 years. However, it is clear with its deep capabilities in content creation and information technology, India remains a potential force in games and simulation. In fact, with its back-end infrastructure prowess, India may become a major force in corporate games, simulations, and MMOG systems, all of which potentially require the skills housed in giants like Infosys and TATA.

The global growth in games has created more open questions of how long the United States may remain in the lead. In general, it is important to realize some underlying truths of the present situation:

- In Korea one of the most popular games of all time is Blizzard's *Starcraft*, a U.S.-made game. In addition EA's *FIFA*, which adopts the Free 2 Play model, is doing very well. U.S. companies are adapting to the new business models slowly but with increasing success. This shows that dominance in gaming can quickly cross borders.
- In China one of the dominant MMOs is Blizzard's *World of Warcraft*, a U.S.-made game. China's games industry studios are dominated by Western interests, although this may eventually change as local companies begin to arise and compete for talent.
- India is far behind in many respects. While expected to catch up, India likely will see much of its first uptake through adoption of popular established games.
- Aside from Brazil, much of South America is populated by very small studios that occasionally do work for larger Western interests.
- Eastern Europe and Russia are more integrated into the existing Western European games industry structure, with only one or two larger Russian-based interests (e.g., 1C, a Russian games publisher) active beyond a regional level.
- The Middle East and Africa are not on the map at all. Middle Eastern investment in university programs for science, information technology, computer science, and the popularity of games with populations skewed so young may provide potential, but so far it is largely unrealized. Pockets of activity have been most visible to the West in the modification of off-the-shelf games for localized rhetorical value. These games by all standards of design are poor but on a level of rhetorical value and innovation are fairly high.
- Despite Japan's dominance in console gaming, Microsoft has made key inroads in the market and Apple has with handhelds. It is rumored that Apple may eventually allow games using its App Store model on its AppleTV device, which also could be a game changer in some respects for the console market. As games increasingly become more about services and social models, U.S. companies may see a new wave of offense versus Japanese dominance. The U.S. comeback in this space has been costly, but should be seen as an example not just of U.S. innovation, but more importantly that seemingly entrenched interests in games can be dislodged.

What is universally true is that games are becoming a major global entertainment and business phenomenon. As tools and techniques easily flow across national boundaries, the ability for new interests to jump into the market is increasing and the effects of that are beginning to take shape. However, the games business at certain echelons is increasingly one of capital and workforce scale, and that allows larger interests wherever based to play significant roles in the face of emboldened local competition. What this shields, however, is whether these increasingly global game concerns dominate with locally produced content and investment or through exported talent and product. It may be that while the names on the door are familiar U.S., European, and Japanese names, they are simply localized talent clusters operating under a familiar name. Talent is the biggest cost in games, and already we are seeing signs of global competitive skew in this regard, especially in Canada with its generous subsidies in Quebec and now Ontario. European interests are exceedingly calling for their own subsidies as well, although the current economic climate may hold some of that back.

As talent becomes more global, and local tastes and coddled companies emerge, the games business will undoubtedly be less U.S. centric in both innovation and execution, but for the near term (10 to 20 years) there is a lot of residual momentum and capability to believe that much of that growth will also involve U.S. interests. More important to watch for is not the overall pendulum swings of business dominance but general capability within local economies that can be sourced by government or third-party interests for nonentertainment application. This is already happening. The question will be one of scale and impact as the serious games and SIM side of the equation is given a boost by the general rising tide of interactive entertainment development in regions such as Asia.

FINAL THOUGHTS

There is great potential for serious application if game developers and users utilize what makes games different than other media: they are interactive, engaging, and fun. The serious application of games will help parties reach young people in this country and internationally. There are great potential applications in learning from games. Mayo (2007) found that game-based presentation of material developed by a first-rate development team is better by a factor of 2 than an outstanding lecturer.

Research continues to posit various such comparative advantages to learning packaged in modern-day computer game form (Gee, 2003; Shaffer, 2008; Collins and Halverson, 2009). Many of the advantages involve key motivational constructs but also new forms of superior pedagogical capability. All of this is more hypothesis than fact while evidence is being gathered in bits and pieces. However, it is worth noting that much of the current belief in games offering comparative advantages for learning and training is grounded in generally posited truths of the effects from multisensory media and the many basic tenants of good learning psychology that games often intrinsically follow (Gee, 2003). Some of the most outwardly daring evangelists of games for teaching come from the viewpoint that games are, in general, naturally more engaging and more attuned to the fundamentals of how people learn than many other means, especially other learning media. Such observations imply that games may one day serve not only as teaching aides but as stand-alone teaching mechanisms that could revolutionize education. While some have posited this notion, it is important to note that a number of leading advocates of games for education in fact see games doing far better among an ecosystem of teachers, mentors, and peers (Shaffer, 2008). This view involves the acceptance of games in various forms of stand-alone teaching but does not acknowledge that it is creating a displacement of mentors, peers, and conventional instructors. Instead, it may be more accurate to describe such ascendency for games as disrupting and realigning the roles of teachers, rather than a simple replacement.

Games are visual and interactive and allow the rapid exploration of potential outcomes in fictional worlds. If we plug in real-world news feeds and data, we can explore potential outcomes in real worlds and revolutionize analysis, perhaps even replacing traditional analysts with online games. If we connect those analytic games up to social media, we can build a "hive mind" of understanding of events happening in real time. Moreover, in the greater context of modeling and simulation, games have the power to facilitate improved simulation of human behavior and group dynamics.

In short, there is more to games than just play. Shared play and experience becomes part of one's culture and perhaps a quite formative part. Games are collaborative and indirectly train large numbers of people with command-and-control experience. In the gaming medium, players get to test hypotheses and learn consequences and then internalize those lessons for later application in life. There is an internalized postplay culture from any long experience with a particular piece of interactive entertainment. With the right understanding of the diverse components and impact of games, the DoD intelligence community and other parties within and outside the industry may better take advantage of this internalized cultural experience of games.

REFERENCES

Published

Abt, Clark C. 1970. *Serious Play*. New York: Viking Press.

Alexander, Leigh. 2009. *WoW* China transition to spell downtime, possible user declines. *Gamasutra*, May 28. Available at http://www.gamasutra.com/php-bin/news_index.php?story=23810. Accessed July 9, 2009.

Apple, Inc. 2009. *Apple Reports Second Quarter Results: Best March Quarter Revenue and Earnings in Apple History*. April 22. Available at http://www.apple.com/pr/library/2009/04/22results.html. Accessed June 4, 2009.

Ashby, Alicia. 2009. Korean game portal profits soar on record virtual goods sales. *Virtual Goods News*, February 16. Available at http://www.virtualgoodsnews.com/2009/02/korean-game-portal-profits-soar-on-record-virtual-goods-sales.html. Accessed July 1, 2009.

Axe, David. 2008. America's army game = brainwashing? *Wired*, January 29. Available at http://www.wired.com/dangerroom/2008/01/army-game-worri/. Accessed July 2, 2009.

Barnes, T., E. Powell, A. Chaffin, and H. Lipford. 2008. Game2Learn: Improving the motivation of CS1 students. Proceedings of the 3rd International Conference on Game Development in Computer Science Education, February 27-March 3, Miami, Florida.

BBC News. 2003. Games at work may be good for you. November 7. Available at http://news.bbc.co.uk/2/hi/technology/3247595.stm. Accessed July 13, 2009.

Blizzard Entertainment, Inc. 2008. World of Warcraft subscriber base reaches 11.5 million worldwide. December 23. Available at http://eu.blizzard.com/en/press/081223.html. Accessed June 5, 2009.

Brumels, Kirk A., Troy Blasius, Tyler Cortright, Daniel Oumedian, and Brent Solberg. 2008. Comparison of efficacy between traditional and video game based balance programs. *Clinical Kinesiology: Journal of the American Kinesiotherapy Association*, Winter. Available at http://findarticles.com/p/articles/mi_6810/is_4_62/ai_n31388272/?tag=content;col1. Accessed June 4, 2009.

Buxbaum, Peter. 2009. Getting serious about games. *Military Training Technology* 14(6):14-17. Available at http://www.military-training-technology.com/mt2-archives/212-mt2-2009-volume-14-issue-6/2191-getting-serious-about-games.html. Accessed January 13, 2010.

Carter, Chip. 2002. Uncle Sam wants you (to play). *St. Petersburg Times*, August 12. Available at http://www.sptimes.com/2002/08/19/Technology/Uncle_Sam_wants_you_t.shtml. Accessed July 9, 2009.

Chen, M. G. 2009. Communication, coordination, and camaraderie in *World of Warcraft. Journal of Games & Culture* 4(1):47-73.

Chowdhry, Amit. 2009. Social network gaming company Zynga pulling in $100 million this year. *Pulse2*, May 3. Available at http://pulse2.com/category/zynga/. Accessed July 9, 2009.

Collins, Allan, and Richard Halverson. 2009. *Rethinking Education in the Age of Technology: The Digital Revolution and Schooling in America*. New York: Teachers College Press.

Cowan, Danny. 2009. Korean government invests $64M into serious games development. *Serious Games Source*, May 19. Available at http://www.seriousgamessource.com/item.php?story=23699. Accessed July 9, 2009.

Ducheneaut, N., N. Yee, E. Nickell, and R. Moore. 2007. The life and death of online gaming communities: A look at guilds in *World of Warcraft*. Pp. 839-848 in *Proceedings of the SIGCHI Conference on Human Factors in Computing Systems*, Bo Begole, Stephen Payne, Elizabeth Churchill, Rob St. Amant, David Gilmore, and Mary Beth Rosson, eds. New York: Association for Computing Machinery.

Durlach, N., Gary Allen, Rudy Darken, Rebecca Lee Garnett, Jack Loomis, Jim Templeman, and Thomas E. von Wiegand. 2000. Virtual environments and the enhancement of spatial behavior: Towards a comprehensive research agenda. *Presence: Teleoperators and Virtual Environments* 9(6):593-615.

Dustman, Robert E., Rita Y. Emmerson, Laurel A. Steinhaus, Donald E. Shearer, and Theodore J. Dustman. 1992. The effects of videogame playing on neuropsychological performance of elderly individuals. *Journal of Gerontology* 47(3):168-171. Available at http://geronj.oxfordjournals.org/cgi/content/abstract/47/3/P168. Accessed July 13, 2009.

Elliott, Jason, Lori Adams, and Amy Bruckman. 2002. No magic bullet: 3D video games in education. *Proceedings of the Fifth International Conference of the Learning Sciences (ICLS)*, Seattle, WA. Available at http://www.cc.gatech.edu/~asb/papers/aquamoose-icls02.pdf.

Falstein, Noah. 2009. Serious games in Europe. *Gamasutra: The Art and Business of Making Games*, March 10. Available at http://www.gamasutra.com/blogs/NoahFalstein/20090310/838/Serious_Games_in_Europe.php. Accessed September 28, 2009.

Feng, Jing, Ian Spence, and Jay Pratt. 2007. Playing video games reduces sex differences in spatial skills. *Psychological Science* 18(10):850-855.

Game Politics. 2007. Iraq War veterans protest America's Army game. *GamePolitics.com*, September 4. Available at http://www.gamepolitics.com/2007/09/04/iraq-war-veterans-protest-americas-army-game. Accessed July 2, 2009.

Gee, John Paul. 2003. *What Video Games Have to Teach Us About Learning*. New York: Palgrave.

Giammattei, Joyce, Glen Blix, Helen Hopp Marshak, Alison Okada Wollitzer, and David J. Pettitt. 2003. Television watching and soft drink consumption: Associations with obesity in 11- to 13-year-old schoolchildren. *Archives of Pediatrics & Adolescent Medicine* 157(9):882-886.

Griffiths, Mark. 2002. The educational benefits of video games. *Education and Health* 20(3):47-51. Available at http://www.sheu.org.uk/pubs/eh203mg.pdf. Accessed July 13, 2009.

Goerger, S., R. Darken, M. Boyd, T. Gagnon, S. Liles, J. Sullivan, and J. Lawson. 1998. Spatial knowledge acquisition from maps and virtual environments in complex architectural spaces. In *Proceedings of the 16th Applied Behavioral Sciences Symposium*. Colorado Springs, CO: U.S. Air Force Academy, pp. 6-10.

Haier, Richard J., Sherif Karama, Leonard Leyba, and Rex E. Jung. 2009. MRI assessment of cortical thickness and functional activity changes in adolescent girls following three months of practice on a visual-spatial task. *BMC Research Notes* 2:174. Available at http://www.mrn.org/latest/brain-imaging-shows-playing-tetris-leads-to-both-brain-efficiency-and-thicker-cortex. Accessed October 23, 2009.

Horswill, I., and M. Novak. 2006. Evolving the artist-technologist. *IEEE Computer* 39(6):62-69.

Jenkins, Henry. 2006. *Fans, Bloggers, and Gamers: Exploring Participatory Culture*. New York: NYU Press.

Jenkins, Henry. 2008. *Convergence Culture: Where Old and New Media Collide*. New York: NYU Press.

Kennedy, Brian. 2002. Uncle Sam wants you (to play this game). *New York Times*, July 11. Available at http://www.nytimes.com/2002/07/11/technology/uncle-sam-wants-you-to-play-this-game.html?sec=&spon=&&scp=1&sq=wardynski&st=cse. Accessed July 2, 2009.

Kibb, Andre. 2007. Brain hand games. *Treo Central*, December 3. Available at http://www.treocentral.com/content/Stories/1437-1.htm. Accessed June 19, 2009.

Martin, Matt. 2009. Nintendo DS ships 100 million units. *GamesIndustry.biz*, March 11. Available at http://www.gamesindustry.biz/articles/nintendo-ds-breaks-100-million-sales. Accessed June 29, 2009.

Mayo, Merrilea. 2007. Games for science and engineering education. *Communications of the ACM* 50(7):30-35.

Mazel, Jacob. 2009. Wii *Fit* sales top 20M worldwide, revenue nearing $2 billion. *VGChartz*, June 16. Available at http://news.vgchartz.com/news.php?id=4018. Accessed June 19, 2009.

McDonald, Marc, Robert Musson, and Ross Smith. 2007. Using productivity games to prevent defects. *The Practical Guide to Defect Prevention*. Microsoft Press. Available at http://my.safaribooksonline.com/9780735622531/ch05. Accessed July 13, 2009.

Medical News Today. 2008. Wii-habilitation: Using video games to heal burns, also using "Guitar Hero" game. July 16. Available at http://www.medicalnewstoday.com/articles/115170.php. Accessed June 5, 2009.

Murphy, Emily C., David Donley, Linda Carson, Irma Ullrich, Justine Vosolo, Chris Mueller, Kim Richison, and Rachel Yeater. 2006. An innovative home-based aerobic exercise intervention improves endothelial function in overweight West Virginia children. *Medicine & Science in Sports & Exercise* 38(5):S571.

Murphy, Emily C. S., Linda Carson, William Neal, Christine Baylis, David Donley, and Rachel Yeater. 2009. Effects of an exercise intervention using *Dance Revolution* on endothelial function and other risk factors in overweight children. *International Journal of Pediatric Obesity* 4(4):205-214. Available at http://www.informaworld.com/smpp/content~db=all~content=a910209496. Accessed July 13, 2009.

Ng, Keane. 2009. Play *Peggle*, lose weight. *The Escapist*, March 25. Available at http://www.escapistmagazine.com/forums/read/7.102302. Accessed July 1, 2009.

NRC (National Research Council). 1997. *Modeling and Simulation: Linking Entertainment and Defense*. Washington, DC: National Academy Press. Available at http://www.nap.edu/catalog.php?record_id=5830.

Okagaki, L., and P. Frensch. 1994. Effects of video game playing on measures of spatial performance: Gender effects in late adolescence. *Journal of Applied Developmental Psychology* 15:33-58.

Parberry, I., T. Roden, and M. G. Kazemzadeh. 2005. Experience with an industry-driven capstone course on game programming. *ACM SIGCSE Bulletin* 37(1):91-95.

Phelps, A., C. Egert, and K. Bierre. 2006. Games first pedagogy: Using games and virtual worlds to enhance programming education. *Journal of Game Development* 1(4):45-64.

Pontin, Jason. 2007. Artificial intelligence, with help from the humans. *New York Times*, March 25. Available at http://www.nytimes.com/2007/03/25/business/yourmoney/25Stream.html. Accessed June 19, 2009.

Rice, John. 2007. Programming a new AquaMOOSE? Virtual real worlds using MellaniuM & Unreal 2. *Educational Games Research* blog, November 25. Available at http://edugamesblog.wordpress.com/2007/11/25/programming-a-new-aquamoose-virtual-real-worlds-using-millanium-unreal-2/. Accessed June 5, 2009.

Satava, Richard M., K. S. Morgan, Hans B. Sieburg, R. Mattheus, and J. P. Christensen. 1995. *Interactive Technology and the New Paradigm for Healthcare*. Amsterdam: IOS Press.

Securities and Exchange Commission. 2009. *Annual Report (Form 10-K) for Activision Blizzard, Inc., for the Year Ended December 31, 2008*. Available at http://investor.activision.com/secfiling.cfm?filingID=1047469-09-2015. Accessed June 29, 2009.

Shaffer, David Williamson. 2008. *How Computer Games Help Children Learn*. New York: Palgrave Macmillan.

Squire, Kurt. 2003. Video games in education. *International Journal of Intelligent Simulations and Gaming* 2(1):49-62.

Subrahmanyam, Kaveri, and Patricia M. Greenfield. 1996. Effect of video game practice on spatial skills in girls and boys. *Interacting with Video*. Patricia Marks Greenfield and Rodney R. Cocking, eds. Santa Barbara, CA: Greenwood Publishing Group. Available at http://books.google.com/books?id=QZjJ9uIj1CcC&printsec=frontcover. Accessed July 13, 2009.

Sung, Kelvin. 2009. Computer games and traditional CS courses. *Communications of the ACM* 52(12):74-78.

Thai, Ann My, David Lowenstein, Dixie Ching, and David Rejeski. 2009. *Game Changer: Investing in Digital Play to Advance Children's Learning and Health*. Report from the Joan Ganz Cooney Center at Sesame Workshop, June. Available at www.joanganzcooneycenter.org/pdf/Game_Changer_FINAL.pdf. Accessed July 9, 2009.

Tychsen, Anders, and Michael Hitchens. 2009. Game time: Modeling and analyzing time in multiplayer and massively multiplayer games. *Journal of Games and Culture* 4(2):170-201.

Wagner, James Au. 2009. The top 10 money-making MMOs of 2008. *Salon.com*, February 1. Available at http://www.salon.com/tech/giga_om/tech_insider/2009/02/01/top_10_money_making_mmos_2008/print.html. Accessed June 19, 2009.

Whitehead, J. 2008. Introduction to game design in the large classroom. *Proceedings of the 3rd International Conference on Game Development in Computer Science Education*. New York: Association for Computing Machinery, pp. 61-65.

Williams, Dmitri. 2006. Groups and goblins: The social and civic impact of online gaming. *Journal of Broadcasting and Electronic Media* 50(4):651-670.

Williams, Dmitri, Nick Yee, and Scott Caplan. 2008. Who Plays, How Much, and Why? Debunking the stereotypical gamer profile. *Journal of Computer-Mediated Communication* 13:993-1018.

Wolz, U., and S. M. Pulimood. 2007. An integrated approach to project management through classic CS III and video game development. *ACM SIGCSE Bulletin* 39(1):322-326.

Wright, Will. 2006. Dream machines. *Wired*, April (14.04). Available at http://www.wired.com/wired/archive/14.04/wright.html. Accessed July 2, 2009.

Yates, Diana. 2008. Strategic video game improves critical cognitive skills in older adults. Press release. University of Illinois News Bureau. Available at http://news.illinois.edu/news/08/1211gamers.html. Accessed July 9, 2009.

Yee, N. 2006. The demographics, motivations, and derived experiences of users of massively multi-user online graphical environments. *Presence: Teleoperators and Virtual Environments* 15:309-329.

Yee, N. 2009. Befriending ogres and wood-elves: Relationship formation and the social architecture of Norrath. *International Journal of Computer Game Research* 9(1). Available at http://gamestudies.org/0901/articles/yee. Accessed June 24, 2009.

Zyda, Michael. 2005. From visual simulation to virtual reality to games. *IEEE Computer* 38(9):25-32.

Zyda, Michael, Victor Lacour, and Chris Swain. 2008. Operating a computer science game degree program. *Proceedings of the 3rd International Conference on Game Development in Computer Science Education*. New York: Association for Computing Machinery, pp. 71-75.

Zyda, Michael, Marc Spraragen, and Balakrishnan Ranganathan. 2009. Testing behavioral models with an online game. *IEEE Computer* 42(4):103-105.

Unpublished

McBride, Dennis. 2009. A conversation. Presentation to the committee on February 27, 2009.

4

Defense Modeling, Simulation, and Games

INTRODUCTION

Modeling, simulation, and games (MS&G) facilitate a broad set of capabilities for analysis focused on defense and national security. The United States uses modeling and simulation (M&S) to advance our understanding of human genetics, weather prediction, crop failure, ballistic performance of new weapons, aerodynamic optimization of fighter aircraft and other warrior, weapons, and information system performance against strategic, operational, and tactical threats and adversaries. Additionally, there has been an increasing dependence on serious games (see Chapter 3) to provide training aids to our soldiers and engaging learning environments to keep our youth interested in education as they step onto a path guaranteeing a continuation of our nation's innovation cycle. Finally, we observe a growing potential for social impact with national security considerations as games become more interconnected and create their own cultures.

However, the United States as an entity is not the only country, group, or nonstate actor looking to leverage the advantages of MS&G to satisfy strategic and tactical goals, whether they be academic, economic, or military in nature. As the basic components of M&S technology (computer architecture, software, and algorithms) become commodity in many respects, widespread use of these technologies will no longer be controlled by a powerful few (Lardinois, 2009; Ohmae, 1995).

Particular disciplines within the realms of MS&G have deeper potential impacts. Scientific modeling and simulation, long driven by advances in computer architecture and a quest for realistic representation and problem solving, is gaining wider penetration and practical use outside its traditional community as games leverage realistic models such as physics engines (Ponder et al., 2003). Furthering this penetration, open-source communities are distributing prepackaged scientific knowledge in a consumable form for reintegration in code bases that can be leveraged by other parties with minimal investment (Tirole and Lerner, 2000; Bonaccorsi and Rossi, 2003). It is only natural that gains realized in the commercial sectors are being redeployed to the defense industry in the form of virtual reality training environments and kinetic and nonkinetic weapons development and deployment.

This chapter will discuss the defense and national security implications of modeling, simulation, and games. In particular, the topics of scientific modeling and simulation, cyber and kinetic warfare, propaganda through games, and war gaming will be explored.

SCIENTIFIC MODELING AND SIMULATION

The increased power of computers arising from faster chips and memory, better bandwidth, better algorithms and architecture, and other improvements described in Chapter 2 means that increased levels of realism, accuracy, and fidelity can be included in models, whether a model is used for science and engineering or for games.

The level of detail included in a simulated model must be optimized for the complexity of the problem, the hardware platform that will be used to solve the model, the level of qualitative versus quantitative accuracy required, and the time one is willing to wait for the solution. As computer speed increases, the accuracy and fidelity of the solution can also increase while the time to solution decreases. In science, recent increases in computing hardware speeds and capabilities mean that more detailed underlying physics (e.g., fluid flow, thermodynamics, molecular interactions) can be considered within models of radar cross sections, airfoils, weather and climate, and materials, respectively, providing a greater level of predictive accuracy. There are particular fields in which huge breakthroughs in simulation capabilities are having a positive effect, as mentioned in the World Technology Evaluation Center report (Oden et al., 2006; WTEC, 2009). Examples include the life sciences and medicine, where the ability to perform simulations of complex biological molecules such as proteins on millisecond timescales with atomistic resolution is now possible.

Verification and validation (V&V) and uncertainty quantification (UQ) are essential elements of modeling and simulation and are critical for proper risk assessment. Despite this, systematic efforts to develop these elements and integrate them into scientific modeling and simulation are not yet prevalent in either academic research programs or industrial research efforts. The 1991 Army Science Board noted that the Army's use of "unvalidated simulations" should end its use of M&S until V&V is addressed. The report claimed the Army had been overly focused on the "pretty graphics" and fast run time of the sim (Lynn et al., 1991). A more recent study (WTEC, 2009) found that the United States leads only marginally in V&V and UQ through the efforts of major Department of Energy programs (ASC/ASCI/PSAAP[1]). Progress in the ability to automate V&V and UQ for scientific simulation would likely show up in the typical academic literature, but may also be developed in-house as proprietary efforts, especially in industrial labs. General methods and strategies developed for V&V and UQ for a given scientific problem are expected to be transferable to some extent to other domains. Rapid progress in this area would give a major competitive edge to the level of predictability of simulation and the accurate assessment of risk. See Chart 4-1 for a representation of this scenario according to the technology warning methodology detailed in Appendix C.

The ability to include physics-based models in simulation can be exploited to increase the "realism" of games through increased detail and more accurate levels of fidelity of the models solved as part of the games program. Examples of real-world intersections between games and M&S are detailed as case studies in Boxes 4-1 and 4-2.

While the open-source movement is seen by many as critical for leveraging scientific, engineering, and business productivity code bases, there is still the inherit danger in free access to information by those with nefarious purposes. In fact, some individuals in the U.S. military, homeland security, and

---

[1] ASC stands for Advanced Simulation and Computing, ASCI is the ASC Accelerated Strategic Computing Initiative, and PSAAP is the Predictive Science Academic Alliance program.

CHART 4-1 Automation of Verification and Validation (V&V) and Uncertainty Quantification (UQ)

Technology	Observables
V&V and UQ are essential elements of modeling and simulation and are critical for proper risk assessment, yet systematic efforts to develop these elements and integrate them into scientific modeling and simulation are not prevalent in either academic research programs or industrial research efforts.	Evidence of progress may be found in: • Published academic research • Industrial lab output Confidence intervals are expected to be provided with a simulation study. Hurricane tracking has reached this level of maturity.

Accessibility	Maturity	Consequence
Level 1[a]	Technology watch	General methods and strategies developed for V&V and UQ for a given scientific problem are expected to be transferable to some extent to other domains. Rapid progress in this area would give a major competitive edge to the level of predictability of simulation and the accurate assessment of risk.

[a]Many V&V and UQ technologies and tools exist for the physical sciences but are not widely used. This will change through customer expectations. Investments are required to mature V&V and UQ for the social sciences and data-driven applications.

BOX 4-1
Case Study 1: Kinetic Kill Physics Serious Game

The intersection of games and scientific simulation is intriguing, particularly in the field of serious games (see Chapter 3 discussion). Given the popularity of games throughout the world, it is likely that a gaming interface to scientific simulation tools would create more users of the tools. If properly designed, it would be possible to perform useful scientific simulations without being an expert in the field.

Given the great number of Internet users searching for unique and perhaps free gaming environments, consider a mythical example of a serious game of two players in an environment of kinetic kill physics. Behind the scenes could be one of many production-capable hydrodynamics codes. The interface presents Player 1 with the design tools, applicable materials, and parameters to develop explosively fired projectiles (EFPs). Player 2 is provided a tank to design multilayer armor based on both modern and evolutionary materials. Unknown to Player 2, the parameters are limited to the possibilities of what is publically known and speculated about Country A's latest deployed tank in Country B. The players design, shoot, and defend. Points are awarded, and players can reach higher levels of distinction by winning against multiple opponents. While two players score points against each other, a nation or nonnation-state collects data on the most effective EFPs. In effect, the Internet is used to optimize EFPs against Country A, free of charge.

In the example above, the hydrocode could be either an open-source code or one constructed by open-source software libraries. Open source is a celebrated trend for many researchers and is helping accelerate all forms of software development. Open source allows for collaboration across the globe, helps eliminate "reinventing the wheel," and facilitates rapid application development. A catalog of many popular open-source projects can be found at http://www.ohloh.net, which claims more than 3 billion lines of code by 200,000 developers. Open source also enables leveling the playing field across the globe for application software.

> **BOX 4-2**
> **Case Study 2: U.S. Asset Assault Simulation**
>
> Another example of exploiting the new wave of Internet capabilities and open-source software would be in planning an assault on U.S. assets. It is not difficult to imagine a rogue group coupling Google Street View with computer-aided design software to provide a working model for an accurate simulation of a kinetic weapon or weapon of mass destruction attack. One can even imagine virtual-world war games providing accurate geometries of U.S. assets. Players across the globe could work in teams to defend or take out assets. Tactics and results could be recorded for future military plans.
>
> By adopting an open-source strategy, countries with small scientific software investments can rapidly catch up to the big players in terms of a software infrastructure, including parallel software capability. One can find open-source biological and chemical software and, perhaps more importantly, the major building blocks for scientific software: finite element libraries, matrix solution libraries, visualization software, and more. A search on "parallel program" results in greater than 200 products. Fortunately, useful simulations for kinetic and nuclear weapons are highly dependent on accurate material behavior models. To date, accurate material behavior models have required a delicate, costly experimental program to provide data to validate theory. These models tend to be considered important intellectual property and are protected as proprietary, export controlled, or classified.
>
> A watch sign for heightened threats in this area would be the appearance of highly accurate and open-source material behavior models or the emergence of robust technologies to accurately link length scales between atomistic models to continuum models. Unfortunately, models for crude but effective distribution of chemical, biological, or radiological agents are easily obtainable in the open literature.
>
> This scenario is represented as technology warning in Chart 4-2.

CHART 4-2 Vulnerability of Physical Assets Given Open-Source Software and Information

Technology	Observables	
Internet-based capabilities using open-source software (e.g., coupling Google Earth with Google SketchUp) for simulation of a kinetic attack). The virtual war game could be run using physics-based modeling of the objects and materials and geolocation similar to the Defense Advanced Research Projects Agency's RealWorld.	Heightened risk may be found given: • The emergence of well-modeled simulations that allow users to adjust the critical coefficients to tune a model and turn a mid-fidelity model into a high-fidelity model rapidly; • The appearance of highly accurate and open-source material behavior models; or • The emergence of technologies to accurately link length scales between atomistic models to continuum models.	
Accessibility	Maturity	Consequence
Level 1	Technology warning	A major competitive edge to the level of predictability of simulation and the accurate assessment of risk. Moreover, increased vulnerability of physical assets.

intelligence communities do find open-source information such as Google Earth to be a threat in the same way that any open-source intelligence can be used to exploit weaknesses in defenses. Access to information will always be creatively used and exploited. This is not to say that there is an overwhelming movement to ban the use of this incredibly helpful technology; rather it is a source of information that needs to be considered controlled by the enemy and can be used against us. Entities that create powerful software that can be readily used to manufacture weapons, artificial intelligence, or other preventable means should carefully consider the impact those advances could bring if openly distributed.

As noted in Box 4-2, a rising threat is the accessibility of open-source software and information concerning U.S. assets. Chart 4-2 further details the possible consequences of this open access.

CYBER AND KINETIC WARFARE

> Practically everything that happens in the real world is mirrored in cyberspace. For national security planners, this includes propaganda, espionage, reconnaissance, targeting, and—to a limited extent—warfare itself. (Geers, 2008)

There is a growing movement to use a wide range of modeling, simulation, and game technologies to advance the state of the art of cyber and kinetic warfare. The existence of mature virtual worlds, in terms of capability and user base, continues to add credence to the notion of cyberspace not only as a medium of information transport but also as a point of organization and even social identity. From within those boundaries we can begin to expect an increase in all forms of activity, including those influenced by modeling, simulation, and especially games. As the capabilities of tools, virtual worlds, and accessibility gain momentum there will be further influence on the way state and nonstate actors wage war. The impact on national security of kinetic warfare is obvious, historic, and dramatic. On the other hand, the migration of espionage, reconnaissance, and propaganda to the cyber realm—enabled by autonomous connections between people and networks (i.e., the Internet)—is still less than 30 years old. In conflicts from Chechnya in 1994 up through the current conflicts in the Middle East, cyber warfare has become an important component and force multiplier of the ground warfare effort (Lewis, 2002; Geers, 2008).

The fields of modeling, simulation, and games have the potential to yield an asymmetric advantage in cyber warfare. Enemies could model cyber capabilities for preplanning of electronic and cyber warfare campaigns. Automation of attackers in this domain is a critical area of interest as distributed computer architectures add more power to thousands of mobile devices. Multiplayer online games, as well as more ephemeral games, can be used to augment the distribution of propaganda or viruses. See Chart 4-3 for a summary of these concerns in the technology warning representation.

Additionally, technology may evolve to a level of realism and interconnection that would allow actual cyber warfare attacks to be launched and controlled from a game environment. The application of artificial intelligence, agent-based simulation, and realistic cyber environments provides the basic building blocks of a computational analytic framework that allows users to explore the space of potential outcomes. Finally, the drive for ever more powerful computer architectures may produce a complete separation from the current trajectory, of ever more capable human-controlled attacks, and change the face of cyber warfare efforts completely, such as through entirely automated cyber warfare campaigns aimed not just at spreading viruses but also intelligently planning and carrying out online attacks and defenses that are integrated with kinetic and psychological operations activities of the real world.

CHART 4-3 Cyber Warfare Through Virtual Worlds

Technology	Observables
The existence of mature virtual worlds, in terms of both capability and user base, continues to add credence to the notion of cyberspace not only as a medium of information transport but also as a point of organization and even social identity.	Games that introduce real connections to real life (e.g., external Internet connectively allowing links to Supervisory Control and Data Acquisition systems), particularly in the form of virtual worlds.

Accessibility	Maturity	Consequence
Level 2	Technology watch	Enemies could model cyber capabilities for preplanning of electronic and cyber warfare campaigns. Multiplayer online games, as well as fun, viral, ephemeral games, can be used to augment the distribution of propaganda or viruses.

Computer Security

While much of the expansive topic of computer security is left out of this report because of the abundance of cyber security publications commissioned by the National Academies (such as NRC, 2009), the committee thinks it is worth mentioning here that computer security also can be modeled, simulated, and then acted on. Agent-based cyber warfare draws extreme parallelism from game engine artificial intelligence (AI) design. Put in simplest terms, information assurance and cyber security as a whole are a game of moves and countermoves between two antagonists where two sides theoretically consist of teams of agents (Hamilton et al., 2002). Therefore, with intelligent algorithms and accurate models of an opposing side's computer systems and information safeguards, an attacking force could launch massively parallel and highly interactive attacks with high confidence of success due to the fidelity of the preceding simulations. The threat of presimulated automated attack is summarized in Chart 4-4.

Finding 4-1: Improvements in and the deployment of agent-based simulation technology—that is, technology that simulates the actions and interactions of autonomous characters and/or systems such that an understanding or a view into the simulated behavior or system can be obtained—as the underpinning of game artifical intelligence systems could be a source of significant vulnerability to the extent that the United States falls behind in this area.

CYBER PROPAGANDA THROUGH GAMES

The Internet has undeniably provided the world's largest international information distribution forum. With information comes disinformation and all manner of propaganda, opinion, saboteurs, and simple mistakes (Chttenden, 2006). Common to nearly all forms of electronically distributed information is the return on investment as it "may be attempted for a fraction of the cost—and risk—of any other information collection or manipulation strategy" (Geers, 2008). The Geers article only scratches the surface of how propaganda can be used through games to affect internal opinions and actions.

CHART 4-4 Presimulated Automated Computer Attack

Technology	Observables
Computer security can also be modeled, simulated, and then acted on. Information assurance and cyber security as a whole are a series of moves and countermoves.	Computer security test-bed environments that allow automated RED-team/BLUE-team testing of large-scale computer networks.

Accessibility	Maturity	Consequence
Level 2	Technology watch	With intelligent algorithms and accurate models of an opposing side's computer systems and information safeguards, an attacking force could launch massively parallel and highly interactive attacks with high confidence of success due to the fidelity of the preceding simulations.

Political Manipulation Through Games on the Internet

Large-scale games create a new venue for political manipulation. While some cases of direct political manipulation of an individual have been popularized by fiction such as *The Manchurian Candidate*, it is the manipulation of large numbers of people that can potentially be accomplished through Internet media. Enabled by the Internet, political messages can be quickly distributed to the world population with little validation beyond the recipients' own social filters. Games, modeling, and simulation present a new dynamic in the use of the Internet as a source of propaganda.

There are three main components that need to exist in order for a game to have an effect on its user base. First, the computer game should be well designed to distribute a political message, either covertly or overtly. In the former case, the game may be more likely to convert new people to an ideology. In the latter, it is more likely to target a population in order to reinforce a message or derive an action from a willing user base. The Serbian resistance and pro-democracy game *A Force More Powerful*, for example, teaches the strategy of nonviolent conflict (Kohler, 2005). The second requirement to make a game more effective is to adequately model the behaviors of the game's actors in order to optimize the messages being sent. Effects of this could be significant if properly orchestrated by a foreign power or commercial entity. A third challenge for game designers is to create a game that is compelling and fun to play in order to keep audiences attracted to it for a substantial enough period of time to successfully convey desired messages.

WAR GAMES

The Evolution of War Games

War games have been used since ancient times to assess the nature of conflict for opposing and threatening forces. These games typically had two purposes: post event to assess the impact of a different decision or choice associated with an action taken (i.e., if we had done this instead of that, would the result have changed?) or planning for a future course of action (if we do this, what do we think will happen?). Chess is one of the oldest surviving ancient war games, dating back some 2,000 years (Dunnigan, 2000).

Regardless of the game, war games share some common attributes: individual pieces (people, equipment, installations) with distinct capabilities, predefined and agreed-on rules of employment; knowledge of their power and effects on other player's pieces; and predefined end states to "win." One side "moves" its game pieces—typically through a preplanned strategy—to achieve a successful objective in an environment that provides barriers to success by an opposing player going through the exact same process. Playing the game provides the player with the ability to think, react, understand, and make better decisions about move and countermove options and to develop better strategic choices about the use of his or her different capabilities in different ways.

The Prussian Army is credited as the first organized army to use true realistic war games as a method to better train, plan, and assess a plan of operations for strengths and weaknesses (Dunnigan, 2000). Moving around small game pieces and equipment on a battlefield map against known or suspected enemy strongholds provided operational insight on the best order of battle in a tactical environment. Massive logistical problems in early World War II led to the development of more complex tabletop simulations for larger scale operations in an effort to solve the age-old problem of getting the right stuff to the right place through the right channels.

As the scale and scope of the battlefield have expanded, the need for higher degrees of scientific modeling has grown exponentially. No longer can a table of game pieces on a small topography effectively capture planning and execution insights into the effects of moving and employing thousands of "pieces" (e.g., people, equipment) across a global map. Understanding the strategic effects of many simultaneous operational battles encompassing thousands of tactical moves has required an even more complex set of rules and attributes—and a lot more scientific computing power. While the initial use of military war games was largely facilitated to study the art of warfare and tactical maneuver, modern war games have taken on a different military value, serving far more:

- As a predictive tool to assess specific combat capabilities and their impacts when employed with other combat capabilities;
- As an assessment of the total number of weapons systems from man to machine and their prioritization, timing, delay, and impact associated with strategic choice among different theaters of operations;
- As a means to study the impacts of attrition due to losses on the battlefield;
- To assess different concepts of operations using existing capabilities;
- For the continued practice and value of "mission rehearsal" that has remained a constant since the initial inception of war gaming; and
- For development of new CONOPs (concept of operations), tactics, and strategies.

Huge, inflexible models were built over time to support the need for simulation decision-level data. The models and their conventional rules began to create artificial barriers to innovative and creative combat rules. As the battlefield grew more complex, the games themselves expanded into massive exercises that required extensive planning, large war game staffs and institutions, and very large and, complex models with increasingly rigid rules. At the same time, actual warfare increasingly took on a more asymmetric tone, where conventional rules of war were increasingly irrelevant as they were replaced by a far more adaptive and rapidly changing enemy.

Over time, traditional war games became missions in themselves, leading to each service developing its own games, staffs, budgets, contractors, and contractor-developed models to support the increasing dependency on these games. Large service war games can take a minimum of one to two years to complete from the start of preparation planning to the final report and out briefs. Each of these massive exercises includes extensive game preparation and manpower support to assess a very limited number of issues (between three and five objectives), culminating in hundreds of players physically traveling to and attending a week-long paper exercise (with limited, if any, use of commercial gaming visualization tools). Participants are given time to execute two to three game moves using primitive models with limited to no adaptability for ingenuity, surprise, and innovation. The staff size alone (both government and contractors) ranges from 20 to 50 people to support game players in the range of several hundred.

The logistical burden of this model of gaming is quite onerous. Once the game has started, creative and unanticipated moves can often stop the play until Game Control inserts instructions prohibiting such moves. Some of these games are supported by decades-old models built and rigidly applied to specific rule sets. These limit the use of any weapons system to a defined method and as a result allow no room for on-the-fly adaptation and creativity. The logistics of game and scenario setup typically cost millions of dollars per single-event game and are not often reusable for subsequent games with different scenarios. In addition, because these games must be played in one physical location, there are travel, housing, and other costs to support anywhere from 200 to 500 players.

In many ways, lack of adaptation to massively multiplayer online war games has created a new "cadre" of the same players for each and every different service game, limiting the input from a broader community whose jobs and missions do not allow attendance and participation in person.

Major Applications of Military War Games

Today's military war games are typically used and valued to accomplish four major efforts:

- Test new CONOPs,
- Assess the impact of resource decisions on future technologies,
- Advocate for resources, and
- Provide next-generation training support.

War Games to Test Concepts of Operations

War games and their results are often used to assess or test new CONOPs as mission rehearsal, training, or "what if" drilling against a specific enemy in a specific region of the world. According to Peter Schnorr and David Perme (2004), the use of simulations for training today's warfighters in today's asymmetrical warfare environments requires investments in advanced technologies that allow more real-time interactive modeling and simulation solutions "on a more unorthodox and spatially complex scale." Schorr and Perme are convinced that the use of intelligent agents and expert systems can aid the development and execution of the simulation events but not without the capability for more tailorable models and rule sets.

An intelligent agent capability might be useful in the collection of data, to create ad hoc queries, and to adjust for lessons learned in each simulation run. Making use of emergent service-based architectures incorporating the flexibility presented by the use of agents and expert systems will allow current and future simulation solutions to behave more unpredictably and erratically, like today's enemy.

War Games to Assess Resource Impact

War games are now used to assess the impact of different resource decisions if investments are made in future technologies and capabilities (some of which do not exist today but have had enough research and development (R&D) investment and science to believe them achievable) with significant adjustments in budget against potential future threats or vulnerabilities. These "future capability" games are often the environment in which to explore the art of the possible and to assess the value of adjusting force structure or technology plans.

These games also provide a mechanism to consider future threats. When building BLUE (cooperative) and RED (adversarial) forces, the game designers must do a thorough review of friendly and threatening capabilities for the next 15 to 20 years. They gather data on actors throughout the world with R&D investments in hostile or lethal capabilities.

Game designers also search for indications of friendly vulnerabilities to develop RED capabilities for their toolbox. There have been attempts to support these games using models, but there have been limited examples of success. More often, these archaic models are criticized for their inability to truly assess the capability impacts where specific technology is still in the "imagine" phase. Instead, the value of these games has become the ability to discover promising new technologies and assess their transformational nature on existing warfare as well as unexpected vulnerabilities that require new capabilities to remedy.

The Air Force's *Future Capabilities Game* is a good example of this improved type of game. In the 2004 game, the Air Force sought ways to refresh its war games methodology to derive better results for less cost and to reduce the logistical game footprint. Game designers sought ways to distribute the game to more than one location to drive down costs and to reach a broader player who, due to mission and job requirements, could not commit to a week-long, full-time war game at a site far from his station (Rolleston, 2009). This 2004 war game reluctantly eliminated rigid models for the first time and gave the game execution a wider free-play environment with a new self-assessment process. The focus this time was not on the cadre of experienced gamers, but on rapidly gathering insights (via statistical data) from a new category of players: future senior leaders and warfighters. This change was made in anticipation that these new players would take what they learned and experienced in this more fluid adaptive environment and apply it to real situations in the future. The one thing lacking was an intelligent agent and adaptive model that operated in real time with each play providing immediate feedback to the gamers.

Both the Air Force's 2004 and 2005 games cited, as a primary objective, "to educate senior Air Force leaders on emerging technologies, innovative system concepts, and emerging Red [adversarial] capabilities" (USAF, 2003). Instead of emphasizing models that would directly determine a winner up-front, time and investment were made in describing, designing, and distributing a comprehensive toolbox (with graphics, pictures, and digital videos of the capability in action and attributes for each discovered new technology, much like today's gaming industry provides for its players). Additionally, more effort was allocated in the postgame period to digitally visualize the war game's outcome for a more game-like experience.

This approach to out briefing the results resulted in better visual and experiential understanding for both senior Air Force leaders and members of the labs and scientific communities who participated.[2] The revised war game was considered to have significantly more impact on these senior Air Force decisions makers, who drew far more from the lessons and threats they could see "in play" than from the previous years' war games. The archaic model-constrained data of the previous games were considered to have inhibited ingenuity and creativity. These small changes in the Air Force game hint at the great

---

[2]Personal communication between Mort Rolleston and Committee Member Allison Hickey on July 8, 2009.

achievement potential possible if all services considered adapting the best practices associated with today's commercial massively multiplayer war games, which are conducted in a virtual environment 24 hours a day with missions constantly impacted, lessons constantly learned, and players adjusting the rules of play at a moment's notice.

War Games to Advocate for Resources

Finally, the use of war games has grown as a tool to influence decisions in national defense reviews like the congressionally mandated Quadrennial Defense Review for the games' ability to demonstrate (with some magnitude of science depending on the service model and its assumptions) one service's capabilities over another in the ever-growing competition for limited defense budgets. Each service uses its preferred models to provide advantage to its forces and weapons systems over another service. Today's military war games play critical roles in service budgets and force structure decision processes.

Throughout the committee's exploration of the breadth of technology drivers in modeling, simulation, and games, there has been a clear demonstration of the benefits that can be derived from widespread adoption of large-scale leaps in commercial development. However, most defense organizations are consumed and trapped by their past success with outdated methods and they largely ignore the leaps in technology, culture, and practice of the growth in other sectors, like the gaming industry. The use of simulation support for operational decision making has been highly desirable but thus far unachievable because the databases that drive the simulations could not be developed and updated fast enough (Schnorr and Perme, 2004).

The U.S. Armed Forces have been slow to adopt automated and intelligent warfare simulation environments. The analogous environments to the pop culture *War Games* movie are barely in existence (e.g., "move," "countermove," "constrained choices"), instead performed through expensive multiyear exercises that simulate relatively few moves between two actor nations. The April 2002 U.S. Army's *Vigilant Warriors* war game held at Carlisle Barracks, in Pennsylvania, for example, included a group of 500 U.S. and allied military and civilian personnel, sequestered in rooms filled with wall-mounted maps, telephones, and computer terminals, conducting "tabletop" exercises, all while focused on using tomorrow's expected military capabilities to mitigate several global crisis scenarios set in the year 2020 (Gilmore, 2002).

Recommendation 4-1: Military war games should exploit the significant growth and lessons of serious games to leverage experiential aspects of large multiplayer joint war games. A more real-time, immediate-feedback exploration environment can then be assessed using rapidly updated algorithms, parameters, and coefficients that reflect behavioral and policy implications. The use of serious games to explore strategy or technology implications can be valuable to strategic and long-range concepts of operations, weapons system acquisition, and threat assessment and response and far more effective in providing constant assessment of new technology and CONOPs opportunities, as well as more real-time threat warnings to those who consistently monitor for these issues. This same virtual sandbox can provide rapid assessment of these capabilities in a more affordable virtual method and reduce the manpower-extensive planning, logistical, and cost requirements typically required by large war games that only occur on a two-year basis.

Creating an accessible war gaming environment with the dynamic virtual world and persistence like *World of Warcraft* would provide a far nimbler and more effective approach to conduct real-time assessments of breaking news on emerging threats and therefore improve our ability to counter these threat potentials far earlier in the process. In addition, virtual war gaming environments can be crafted on levels from strategic to tactical and across the full spectrum of military response. These games are used today to train and test tactics, techniques, and procedures (TTPs) and as effective tools for mission rehearsal, cultural development and experience, and familiarization training.

War Games to Provide Next-Generation Training Support

A particular nexus of MS&G is in the form of future training environments.[3] While modeling and simulation have historically focused more on the accuracy of models while the games industry has focused on the user experience, the two properties are not mutually exclusive in many fields, especially training.

As an extension to education, leveraging accurate (validated) computer games has the potential to help improve all aspects of U.S. competitive advantage from medical, basic educational, vocational, and military skills training. While nearly all studies surveyed by the committee mentioned a continued lack of in-depth or systematic studies to fully evaluate the effectiveness of video games in training, many have seen an increased level of participation on the part of the students (Annetta et al., 2009; Carnevale, 2005). In fact, the level of interest from the educational community as a whole shows potential that continued interest in the promise of improved training and education will continue to fund research into analyzing the effectiveness of education (skill acquisition through games is covered further in Chapter 3). However, as indicated by the 1991 Army Science Board survey of M&S initiatives, those investing in the effectiveness of games need to be aware of the validity of the models they are built on (Lynn et al., 1991). Educational games may provide great benefits, but should not be created at the expense of the lesson content.

Conversely, games can be used to create the data foundation of future scientific models. As discussed by Orkin and Roy (2007), the use of games in analyzing social behavior in scenario settings is an incredibly powerful tool to help understand human behavior in certain situations. Particularly, the massively distributable nature of electronic games combined with the ease of their "instrumentation" allows for a degree of study not efficiently accomplishable in the real world. As shown in Figure 4-1, the game *Zero Hour: America's Medic* is designed to teach first responders how to react to large-scale disaster scenarios. This training tool could alternatively be designed and instrumented in a manner that would allow researchers to profile how humans actually behave in such situations. This feedback loop could be a critical step in increasing the efficacy of computer-game-based social models. This discussion and its associated technology watch are summarized in Chart 4-5.

Combining the concepts of games for training and for the introspective use of modeling behavior, some researchers are working on using AI systems to augment interactive media offerings in education (Moreno-Ger et al., 2007). IMS's Learning Design specification is designed to learn from its users' interaction with the system to change its game offerings to the student as well as the game interactions

---

[3]The committee was directed by the study's sponsor to not produce an in-depth report on modeling and simulation for training military forces since this has been a topic of many previous reports. For additional information, see *Impact of Advanced Distributed Simulation on Readiness, Training, and Prototyping*, 1993, Defense Science Board, Office of the Undersecretary of Defense for Acquisition, Technology, and Logistics, Washington, DC, January, available online at http://www.acq.osd.mil/dsb/reports/srp.pdf); *Modeling and Simulation: Linking Entertainment and Defense,* 1997, NRC, National Academy Press, Washington, DC, available online at http://www.nap.edu/catalog.php?record_id=5830); and *Advanced Modeling and Simulation for Analyzing Combat Concepts in the 21st Century,* 1999, DSB, Office of the Undersecretary of Defense for Acquisition, Technology, and Logistics, Washington, DC, May, available online at http://www.acq.osd.mil/dsb/reports/advancedmodeling.pdf. Web sites last accessed September 21, 2009.

FIGURE 4-1 Screen shot from the game *Zero Hour: America's Medic*, which allows first responders to role play various disaster scenarios, such as an earthquake, a bombing, or a sarin gas attack. SOURCE: Image courtesy of Virtual Heroes, a Division of Applied Research Associates, Inc.

themselves based on the interpreted preferences of the student. Taking this concept into the future, the system would ideally be used to isolate the most effective ways to train a subject.

U.S. leaders and industry should consider the benefits that can come from proper use of serious games technology. However, it should be understood that this technology is not likely to stay with the United States, and will likely even be developed elsewhere in some arenas. Already, the Chinese People's Liberation Army has begun to embrace its own "virtual training systems and tools" (Mulvenon, 2008).

Enhanced Military Simulation

Despite the lack of widespread games technology used directly in war games, there has been increased use of actual games by the military. One of the most compelling cases for game-based aids to the military is for basic skill and situational training. While games might never be a true replacement for acquiring skill through physical experience, they might very well function as force multipliers for training—helping to overcome the physical bandwidth limitations or steep costs of many live training facilities. Bandwidth constraints, particularly in the tactical battlespace, are a factor that must be addressed if training and planning games are to be deployed to the field. Cooper's law permits a degree of optimism about the future, but it might be worthwhile to monitor technological progress in this spe-

CHART 4-5 Cultural and Behavioral Modeling Through Virtual Environments

Technology	Observables
Games, training environments, and virtual worlds may be specifically instrumented to log data about how people (or a particular culture) respond in certain scenarios or situations. Were adversaries to obtain such capabilities, there might be ramifications for psychological operations (PSYOPS). These environments also present a new dynamic in the use of the Internet as a source of propaganda. Effects of this could be significant if properly orchestrated by a foreign power or commercial entity.	To detect whether games and other virtual environments are becoming a threat, watch for: • Mass investment in widespread behavioral modeling through game analysis; • Research in using games for cultural analysis; and • Games and other social media that are created to appeal to particular demographics while promoting an agenda.

Accessibility	Maturity	Consequence
Level 2	Technology watch	Effective cultural and behavioral modeling through games and virtual environments could be used to profile the effectiveness of planned terror campaigns—a potential PSYOPS concern—and present opportunities for political manipulation.

cific domain, particularly as it pertains to embedded training, distributed planning, and command and control. Bandwidth constraints translate to having to deploy rendering capability (graphical or game play) further on the edges of the network. Any organization or state has the potential to effectively offer richer experiences that permeate the population as a result of increased bandwidth (e.g., South Korea).

The U.S. Army has demonstrated the highest interest of all the services in leveraging commercial gaming capabilities, through its new Games for Training program, and appears to take this approach seriously, as supported by the funding of different initiatives (Turse, 2003). In fact, the 2008 release of the Army Field Manual specifically called out video games as "serious training tools" (McLaughlin, 2009). Recent Department of Defense (DoD)-sponsored game programs include:

- *America's Army* (2002);
- *DARWARS Ambush* (2003);
- *Full Spectrum Warrior* (2004);
- *Stability Operations: Winning the Peace* (2005);
- *Virtual Battlespace 2* (*VBS2*; 2007);
- *Game After Ambush* (2009); and
- *RealWorld* (2009).

The U.S. Army is prepared to invest $50 million in video games to train soldiers for combat. The funds will be used over five years beginning in 2010, as part of its Games for Training program (Jenkins, 2008).

The Army and other services are investigating all forms of interactive media and games to augment a wide range of goals, from engaging scenarios to help service members cope with combat stress, to constructing immersive environments aimed at providing a wide range of realistic scenarios for tactical engagement and role play (see Figure 4-2 for an example).

Despite increased optimism about games, a number of questions have yet to be answered, such as:

- How can enemies use games to learn our tactics and play against us?
- Can virtual worlds really teach people to be better shooters or small ambush squads?
- Can games be used to train and adapt rapidly to CONOPs capabilities and to leverage command and control training capabilities for joint and coalition partners?
- Can online or virtual reality platforms provide for a massive increase in militia-style reserve forces? (This question is of particular concern for guerilla tactics.)

Massively multiplayer online role-playing games (MMORPGs) may provide insights into effective new campaigns for terror, PSYOPS, and mass reaction to certain catalysts that can be developed in online environments to provide rapid evaluation of new TTPs, as indicated in the "Global Industry Trends" section of Chapter 3.

Finding 4-2: While the United States continues to leverage superior training as part of its ability to maintain asymmetric advantages over potential adversaries, these same potential adversaries may develop the ability to train and adapt CONOPs based on prolific access to Western game genres and actors.

FIGURE 4-2 Scenario from the University of Southern California Institute for Creative Technologies' Flatworld prototype for tactical engagement and role play. SOURCE: Image courtesy of the University of Southern California Institute for Creative Technologies. Photograph by Bradley Newman.

Finding 4-3: In the global war on terrorism, American forces have frequently prevailed in direct-fire exchanges, often attributed to better squad coordination and training—a skill commercial multiplayer gamers practice and develop virtually and routinely. Though it is not expected that this advantage will be maintained when engaged with a sophisticated potential adversary in the field, the United States is at risk of losing its advantage due to the advanced training environments and distributed nature of simulation and online gaming either currently or in the near future.

Recommendation 4-2: DoD should strongly consider migrating at least one of its Title 10 war games to the emerging architectures of the commercial gaming industry.

The Air Force's *Future Capabilities* game is a potential candidate to make this leap in order to leverage flexible game play, minimize logistical footprint, and optimize mass player participation across the joint forces.

FINAL THOUGHTS

In recent decades the United States has had a demonstrated lead in technical and computational innovation. However, the Internet has created an environment for open innovation, bringing disparate groups together, democratizing access to state-of-the-art technologies, and providing for different approaches and views on problem solving. As a result, individuals or small groups now have tools to exert influence. The Internet is having a significant impact on culture and moreover is accelerating technology development, communication, and productivity. Social change itself has been accelerated through globalization of attitudes, values, fashions, lifestyles, and languages (Mack, 2009). Lastly, the Internet is accelerating the pace of business. Those who can adapt are more productive, connected, and prepared.

The dynamics of which country is leading in cyber technology are also surprising. Currently, China has more Internet users than any other country; the United States is second. However, the United States is thirteenth in terms of Internet penetration (as defined by Internet connections per resident), and U.S. broadband penetration is beginning to slow and plateau, while in Indonesia and Latin America it is expected to show high relative growth (Mack, 2009).

Take-away Warnings

The United States is not alone in its pursuit of MS&G excellence. Many nations are actively investing in the next generation of computational power. While active research in fields such as game artificial intelligence, the creation of algorithms, and future hardware platforms is often benign, the capability for new advances in warfare is also being created. The basic scientific knowledge of ballistic penetrators is being packaged into open-source academic software and creating the game AI's that can be reused as smart agent automation for cyber security attacks.

Adversaries may leapfrog U.S. capabilities as we continue to rely on mid-1960s tried-and-true, noncomputational analysis due to institutional and political inertia, exemplified by the growing divide between rapidly accelerating cyber technology and U.S. executive leadership expertise in the area. Some suggest that we are already exiting the information age and are now moving into the conceptual age. Yet some senior leaders have yet to adopt e-mail or use the Web. This knowledge was exploited in the 2008 U.S. presidential race when candidate Senator John McCain was cast as out of touch with technology after admitting to not using e-mail. In the 2007 U.S. Air Force Future Capabilities Game, one of the findings was that "players found it difficult to develop/integrate cyberspace into campaign plans due to . . . understanding of capabilities and how to integrate with other domains" (Rolleston, 2009).

Many government agencies and contractors have very strict rules for Internet use at work. Some have no Internet access. One must question if a cost-benefit study has been conducted by U.S. agencies on Internet use. On the one hand, inappropriate computer use and viruses might be eliminated, but it is difficult to understand how these institutions can stay on the leading edge without learning and experiencing where the rest of the world is headed. Many enlightened companies now encourage their employees to invest time in learning new cyber tools (e.g., *Virtual Worlds*; Anderson, 2009). At the same time, an obstacle to incorporating games and simulations into the government and military is resistance to what are perceived as "toys." The pursuit of fidelity for fidelity's sake, the stipulation to avoid "negative training," and the lack of understanding of what could be used for focused training hamper adoption of these technologies.

Take-away Opportunities

Despite challenges, the United States is in a position to take advantage of exciting new applications in the field of MS&G. As one of the world leaders in the creation of serious and nonserious games, our nation possesses a plethora of the talent needed to leverage the full potential to train our troops and educate our population and to do so in a way that may augment or complement traditional styles of learning. Additionally, environments can be created in which to experiment with new ideas and concepts in a fashion that will continue to outpace existing forms of manufacture and development. These revolutions will not happen without proper resourcing, and U.S. leadership must make sure that hurdles to success are removed and innovation in these areas is encouraged.

REFERENCES

Published

Annetta, Leonard D., James Minogue, Shawn Y. Holmes, and Meng-Tzu Cheng. 2009. Investigating the impact of video games on high school students' engagement and learning about genetics. *Computers & Education* 53(1):74-85.

Bonaccorsi, Andrea, and Cristina Rossi. 2003. Why open source software can succeed. *Research Policy* 32(7):1243-1258.

Carnevale, D. 2005. Run a class like a game show: "Clickers" keep students involved. *Chronicle of Higher Education* 51(42):B3.

Chttenden, Maurice. 2006. Comedy of errors hits the world of Wikipedia. *The Sunday Times*, February 12. Available at: http://www.timesonline.co.uk/tol/news/uk/article730025.ece. Accessed June 24, 2009.

Dunnigan, James F. 2000. *Wargames Handbook, Third Edition: How to Play and Design Commercial and Professional Wargames*. Lincoln, NE: IUniverse.

Geers, Kenneth. 2008. Cyberspace and the changing nature of warfare. *SC Magazine*, August 27. Available at http://www.scmagazineus.com/Cyberspace-and-the-changing-nature-of-warfare/article/115929/. Accessed June 25, 2009.

Gilmore, Gerry J. 2002. Army war games provide azimuth for DoD's future force. News Release, American Forces Press Service, April 29. Available at http://www.defenselink.mil/news/newsarticle.aspx?id=44118. Accessed June 25, 2009.

Hamilton, Samuel N., Wendy L. Miller, Allen Ott, and O. Sami Saydjari. 2002. The role of game theory in information warfare. In *Proceedings of Fourth Information Survivability Workshop*. Available at http://www.cyberdefenseagency.com/publications/The_Role_of_Game_Theory_in_Information_Warfare.pdf. Accessed June 25, 2009.

Jenkins, David. 2008. Report: U.S. Army invests $50M in training games. *Gamesutra News*, November 24. Available at http://www.gamasutra.com/news/serious-games/?story=21237. Accessed June 9, 2009.

Kohler, Chris. 2005. Sir, the gamers are revolting! *Wired*, October 27. Available at http://www.wired.com/gaming/gamingreviews/news/2005/10/69372. Accessed July 13, 2009.

Lardinois, F. 2009. Wolfram|Alpha: Our first impressions. *ReadWriteWeb*, April 25. Available at http://www.readwriteweb.com/archives/wolframalpha_our_first_impressions.php. Accessed June 25, 2009.

Lewis, J. A. 2002. Assessing the risks of cyber terrorism, cyber war, and other cyber threats. Washington, DC: Center for Strategic and International Studies. Available at http://www.steptoe.com/publications/231a.pdf. Accessed June 25, 2009.

Lynn, Verne L., Peter Cherry, Edward Brady, Paul Droulihet, Jr., and William Evers. 1991. *Army Science Board 1991 Summer Study—Army Simulation Strategy*. Available at http://www.dtic.mil/srch/doc?collection=t3&id=ADA250382. Accessed July 13, 2009.

McLaughlin, Seth. 2009. U.S. Army expands use of video games for training. *World Politics Review*. Available at http://www.worldpoliticsreview.com/article.aspx?id=3724. Accessed June 25, 2009.

Moreno-Ger, Pablo, Daniel Burgos, José Luis Sierra, and Baltasar Fernández-Manjón. 2007. A game-based adaptive unit of learning with IMS learning design and <e-Adventure>. In *Creating New Learning Experiences on a Global Scale: Second European Conference on Technology Enhanced Learning*, Erik Duval, Ralf Klamma, and Martin Wolpers, Eds. Berlin and Heidelberg: Springer-Verlag, pp. 247-261.

Mulvenon, James. 2008. "True is false, false is true, virtual is reality, reality is virtual": Technology and simulation in the Chinese military training revolution. *The "People" in the PLA: Recruitment, Training, and Education in China's Military*. Roy Kamphausen, Andrew Scobell, and Travis Tanner, Eds. Carlisle, PA: Strategic Studies Institute of the U.S. Army War College, pp. 49-90.

NRC (National Research Council). 2009. *Technology, Policy, Law, and Ethics Regarding U.S. Acquisition and Use of Cyberattack Capabilities*. Washington, DC: The National Academies Press.

Oden, J. Tinsley, Ted Belytschko, Jacob Fish, Thomas J. R. Hughes, Chris Johnson, David Keyes, Alan Laub, Linda Petzold, David Srolovitz, and Sidney Yip. 2006. *Revolutionizing Engineering Science Through Simulation: A Report of the National Science Foundation Blue Ribbon Panel on Simulation-Based Engineering Science*. Arlington, VA: National Science Foundation. Available at http://www.nsf.gov/pubs/reports/sbes_final_report.pdf. Accessed July 1, 2009.

Ohmae, Kenichi. 1995. *The End of the Nation State: The Rise of Regional Economies*. New York: Simon and Schuster.

Orkin, Jeff, and Deb Roy. 2007. The restaurant game: Learning social behavior and language from thousands of players online. *Journal of Game Development* 3(1):39-60.

Ponder, Michael, George Papagiannakis, Tom Molet, Nadia Magnenat-Thalmann, and Daniel Thalmann. 2003. VHD++ development framework: Towards extendible, component based VR/AR simulation engine featuring advanced virtual character technologies. *Computer Graphics International Proceedings*, pp. 96-104. Available at http://vrlab.epfl.ch/Publications/pdf/Ponder_and_al_CGI_2003.pdf. Accessed January 13, 2010.

Schnorr, Peter, and David Perme. 2004. Improving asymmetric threat representations through the use of intelligent agent technology. Presented at the Spring Simulation Interoperability Workshop, Orlando, FL, Simulation Interoperability Standards Organization. Available at http://www.sisostds.org/index.php?tg=fileman&idx=get&id=2&gr=Y&path=Simulation+Interoperability+Workshops%2F2004+Spring+SIW%2F2004+Spring+SIW+Papers+and+Presentations&file=04S-SIW-063.pdf. Accessed June 25, 2009.

Tirole, Jean, and Josh Lerner. 2000. The simple economics of open source. HBS Finance Working Paper No. 00-059. Available at http://ssrn.com/abstract=224008. Accessed June 25, 2009.

Turse, Nick. 2003. Bringing the war home: The new military-industrial-entertainment complex at war and play. *TomDispatch*, October 17. Available at http://www.commondreams.org/views03/1017-09.htm. Accessed July 13, 2009.

USAF (United States Air Force). 2003. *FY03 Wargaming Assessment Report*. Report written by the Air Force Research Laboratory Information Directorate, Rome, NY. Available at http://www.dtic.mil/cgi-bin/GetTRDoc?AD=ADA420838&Location=U2&doc=GetTRDoc.pdf. Accessed January 13, 2010.

WTEC (World Technology Evaluation Center). 2009. Sharon C. Glotzer, Kim Sangtae, and Peter T. Cummings, Abhijit Deshmukh, Martin Head-Gordon, George Karniadakis, Linda Petzold, Celeste Sagui, and Masanobu Shinozuka, panel members. *International Assessment of Simulation-Based Engineering and Science*. Baltimore, MD: WTEC. Available at http://www.wtec.org/sbes. Accessed June 25, 2009.

Unpublished

Anderson, Laura. 2009. Computing technology landscape and directions: Impact on simulation, modelling, and gaming. Presentation to the committee, January 12, 2009.

Mack, Timothy. 2009. Gaming and Culture: Impact of the Internet. Presentation to the committee, February 26, 2009.

Rolleston, Mort. 2009. Future capabilities game 07 results. Presentation to the committee, February 26, 2009.

Appendixes

Appendix A

Biographical Sketches of Committee Members

Delores M. Etter (NAE), *Co-chair*, joined the electrical engineering faculty at Southern Methodist University on June 2, 2008. She holds the Texas Instruments Distinguished Chair in Engineering Education and is director of the Caruth Institute for Engineering Education. Her research interests include digital signal processing and biometric signal processing, with an emphasis on identification using iris recognition. She has written a number of textbooks on computer languages and software engineering. Dr. Etter previously held the Office of Naval Research Distinguished Chair in the electrical/computer engineering department at the U.S. Naval Academy, where she was a faculty member from 2001 to 2008. She was formerly a member of the electrical/computer engineering departments at the University of Colorado, Boulder (1990-1998), and the University of New Mexico (1980-1989). She was also a visiting professor in the Information Systems Laboratory at Stanford University (1983-1984). Dr. Etter has held two senior executive positions in the Department of Defense. As assistant secretary of the navy for research, development, and acquisition (2005-2007), Dr. Etter was the senior acquisition executive for the Navy and the Marine Corps. She also served as the deputy under secretary of defense for science and technology (1998-2001). Dr. Etter is a former member of the National Science Board and the Defense Science Board. She is a fellow of the Institute of Electrical and Electronics Engineers, the American Association for the Advancement of Science, and the American Society for Engineering Education. Dr. Etter has received the Department of the Navy's Distinguished Public Service Award, the Secretary of Defense's Outstanding Public Service Medal, and the Department of Defense's Distinguished Public Service Medal.

Robert J. Hermann (NAE), *Co-chair,* is a senior partner at Global Technology Partners, LLC. He retired as senior vice president for science and technology of United Technologies Corporation in March 1998. He is a former director of the Defense Department's National Reconnaissance Office and a former senior official at the National Security Agency. Dr. Hermann served as a member of the President's Foreign Intelligence Advisory Board during the Clinton administration. At United Technologies he was responsible for assuring the development of technical resources and full exploitation of science and technology. He was also responsible for the United Technologies research center. Dr. Hermann joined the company in 1982 as vice-president of systems technology in the electronics sector and later served in a series of assignments in the defense and space systems groups prior to being named vice-president of science and technology. Dr. Hermann concluded his tenure as immediate past chairman of the American National Standards Institute (ANSI) Board of Directors at the end of 2002 following a two-year term; he had served as chairman of the ANSI Board of Directors during 1999-2000 and as a member of the ANSI board since 1993. Dr. Hermann continues to serve as a senior partner of Global Technology Partners, LLC, which specializes in investments in technology, defense, aerospace, and related businesses worldwide. Prior to joining UTC, Dr. Hermann served 20 years with the National Security Agency, with assignments in research and development, operations, and NATO. In 1977 he was appointed principal deputy assistant secretary of defense for communications, command, control, and intelligence. In 1979 he was named assistant secretary of the Air Force for research, development, and logistics and in parallel was director of the National Reconnaissance Office. He received B.S., M.S., and Ph.D. degrees in electrical engineering from Iowa State University. Dr. Hermann was chosen for this committee for his expertise in military applications and the intelligence community.

Brian Ballard is the chief technology officer of Ares Systems Group, where he is involved in the development of emerging networking and embedded systems technologies for intelligence, surveillance, and reconnaissance systems and applications in government and military. He is a highly experienced professional in the field of national intelligence systems and computer engineering. Employed for more than 10 years with the National Security Agency (NSA), Mr. Ballard has dealt with all forms of data collection, dissemination, processing, and visualization. As a field operations officer at the NSA, he was a member and team leader in the Office of Target Reconnaissance and Surveillance. He also worked for five years as a global network vulnerabilities analyst. Mr. Ballard holds an M.S. in electrical and computer engineering and a B.S. in electrical and computer engineering from Carnegie Mellon University. He is currently studying for an M.S. in technology management and an M.B.A. at the University of Maryland, College Park.

Pierre Chao is founder and managing partner at Renaissance Strategic Advisors. He is also a senior associate nonresident at the Center for Strategic and International Studies (CSIS). Previously he was a senior fellow and director of defense-industrial initiatives at CSIS from 2003 to 2007. Prior to joining CSIS, Mr. Chao was a managing director and senior aerospace/defense analyst at Credit Suisse First Boston (1999-2003), where he was responsible for following the U.S. and global aerospace/defense industry. From 1995 to 1999 he was senior aerospace/defense analyst at Morgan Stanley Dean Witter. He also served as senior industry analyst at Smith Barney in 1994 and as a director at JSA International, a management consulting firm focused on the aerospace/defense industry. Mr. Chao was also a co-founder of JSA Research, an equity research boutique specializing in the aerospace/defense industry. In 1988-1990 he worked for Prudential-Bache Capital Funding as a mergers and acquisitions banker focusing on aerospace/defense. In 2000, Mr. Chao was appointed to the Presidential Commission on Offsets in International Trade. He is also a guest lecturer at the National Defense University and the Defense Acquisition University. He earned dual bachelor of science degrees in political science and management science from the Massachusetts Institute of Technology.

Robert Gehorsam is president of Forterra Systems, Inc., and has more than 25 years of management experience in the online games and entertainment world. In recent years he has applied his experience to leveraging commercial games technology for national security and enterprise applications. Mr. Gehorsam has participated in several government studies on this subject and speaks frequently on the topic. Before joining Forterra Systems, he served as senior vice president of programming and production at Viacom's CBS Internet Group and previously as senior vice president for programming and production at Sony Online Entertainment. From 1985 to 1992 he founded and led Prodigy's games and educational divisions, launching the first large-scale subscription-based sports simulations and nearly two dozen other online titles. He has held management positions at Scholastic Inc. and has consulted on entertainment and broadband strategy for Microsoft, America Online, CNET, Ziff-Davis, and Children's Television Workshop. Earlier in his career he was an editor at Simon & Schuster, where he helped found its electronic publishing division.

Sharon C. Glotzer is the Stuart W. Churchill collegiate professor of chemical engineering and a professor of materials science and engineering at the University of Michigan, Ann Arbor, and director of research computing in the College of Engineering. She also holds faculty appointments in physics, applied physics, and macromolecular science and engineering. She received a B.S. in physics from the University of California, Los Angeles and a Ph.D. in physics from Boston University. Earlier she worked at the National Institute of Standards and Technology. Her research focuses on computational nanoscience and simulation of soft matter, self-assembly and materials design, and computational science and engineering and is sponsored by the Department of Defense, Department of Energy, the National Science Foundation (NSF), and the J. S. McDonnell Foundation. Dr. Glotzer is a fellow of the American Physical Society. In 2009 she was awarded a National Security Science and Engineering Faculty Fellowship from the Department of Defense that provides substantial long-term funding for her research. She has served on the National Academies' Solid State Sciences Committee; the Technology Warning and Surprise Study Committee; the Biomolecular Materials and Processes Study Committee; and the Standing Committee on Technology Insight—Gauge, Evaluate, and Review. She is involved in roadmapping activities on computational science and engineering, including chairing a pan-agency-sponsored/international assessment of research and development in simulation-based engineering and science (SBE&S), and co-chairing a research directions roadmapping effort in SBE&S for NSF. She is also co-founding director of a new Virtual School of Computational Science and Engineering under the auspices of the NSF-funded Blue Waters Petascale Computing Project at the National Center for Supercomputing Applications.

J. C. Herz is a technologist with a background in biological systems and computer games design. Her specialty is massively multiplayer systems that leverage social network effects, whether on the Web, mobile devices, or exotic high-end or grubby low-end hardware. She currently serves as a White House special consultant to the Office of the Secretary of Defense (OSD). Her current and past defense projects range from aerospace systems to a computer-game-derived interface for next-generation unmanned air systems. Ms. Herz is one of three coauthors of OSD's open technology development roadmap. She serves on the Federal Advisory Committee for the National Science Foundation's education directorate. In that capacity she is helping NSF harness emerging technologies to drive U.S. competitiveness in math and science. She was a member of the National Research Council's Committee on IT and Creative Practice and is currently a fellow of Columbia University's American Assembly, where she is on the leadership team of the assembly's Next Generation Project. In 2002 she was designated a Global Leader for Tomorrow by the World Economic Forum. She is a member of the Global Business Network and is a founding member of the Institute of Electrical & Electronics Engineers Task Force on Game Technologies. Ms. Herz is a term member of the Council on Foreign Relations and is on the advisory board of Carnegie Mellon's ETC Press. She graduated from Harvard University with a B.A. in biology and environmental studies, magna cum laude, in 1993. She is the author of two books, *Surfing on the Internet* (Little

Brown, 1994) and *Joystick Nation: How Videogames Ate Our Quarters, Won Our Hearts, and Rewired Our Minds* (Little Brown, 1997). As a *New York Times* columnist, Ms. Herz published 100 essays on the grammar and syntax of game design from 1998 to 2000. She has also contributed to Esther Dyson's *Release 1.0*, *Rolling Stone*, *Wired*, *GQ*, and the *Calgary Philatelist*.

Allison A. Hickey is an executive strategic planner with Accenture National Security Services (ANSS) and is currently completing service as an advisor to the Defense Science Board for Joint Net Centric Interoperability. She has over 28 years of experience in government, military, and intelligence sectors at national, department, state, and agency levels, including 16 years of "inside the beltway" force structure, policy, and strategy assessment leadership. Brig. Gen. (Ret.) Hickey specializes in cross-cutting, multicultural, and multilevel leadership and management, strategic planning, and policy formulation and deployment. She has extensive experience in sensitive and high-impact research; war gaming and simulation models; corporate programming and budgeting processes; operational and personnel readiness issues resolution; business process reengineering; human capital management strategies; and identification, development, and assessment of new technology requirements. Prior to working with ANSS, she served in the Air Force for 27 years. After 10 years as an Air Force pilot and aircraft commander, she spent her last 15 years in positions that provided advice to chief executive officer levels on analysis and assessment of plans through enterprise-wide war gaming and simulation. She led the development and execution of the Air Force's Title 10 war game, *Futures Game* (the first to use distributed gaming teams), which assessed the long-range investment potential of many new technologies under different political constraints. Her years of routine engagement with a wide variety of security, intelligence and military agencies, and the press have resulted in extensive knowledge of media and legislative affairs. She serves on the U.S. Air Force Academy's Association of Graduates Board of Directors and is an advisor to the AcademyWomen's Board of Directors.

Charles Hudson is vice president for business development at Serious Business, a leading developer of social games for social networks. Previously, he was senior director for business development at Gaia Interactive, a social networking and gaming Web site marketed to teenagers. Before joining Gaia, Mr. Hudson was a product manager for IronPort Systems, a leading provider of antispam hardware appliances that was acquired by Cisco Systems for $830 million in 2007. He then worked in New Business Development at Google, where he focused on new partnership opportunities for early-stage products in the advertising, mobile, and e-commerce markets. He spent several years working at In-Q-Tel, a strategic venture capital group for the Central Intelligence Agency. At In-Q-Tel he focused on identifying investment opportunities that could deliver significant value to the CIA and the commercial market in both the short term and long term. Mr. Hudson holds an M.B.A. from the Stanford Graduate School of Business and a B.A. in economics and Spanish from Stanford University.

James Peery is director of the Computation, Computers, Information, and Mathematics Center at Sandia National Laboratories in Albuquerque, New Mexico. In this role he is responsible for research and development activities in high-performance computing. This center contains the Computer Science Research Institute and the Institute for Advanced Architectures and Algorithms. In addition, Dr. Peery is program director of the National Nuclear Security Administrations (NNSA) Advanced Simulation and Computing (ASC) Program at Sandia. Dr. Peery is also chair of the Computer and Informational Science Research Foundation and the Enable Predictive Simulation Laboratory Directed Research and Development Committee. Prior to returning to Sandia, Dr. Peery worked at Los Alamos National Laboratories in the positions of hydrodynamic experiments division leader, principal deputy associate director of the nuclear weapons program, and program director of (NNSA's ASC Program. Before joining Los Alamos National Laboratory, he worked at Sandia National Laboratories where he led the computational solid mechanics and structural dynamics department and computation physics department. During his career, Dr. Peery has been responsible for the development of state-of-the-art, massively parallel computational tools in the fields of high-energy density physics, shock physics, transient dynamics, quasi-statics, nonlinear implicit dynamics, and structural dynamics. His major research areas are in Arbitrary Lagrangian Eulerian algorithms and parallel algorithms, on which he has published over 50 papers. As part of the SALINAS team, he was awarded the 2002 Gordon Bell Award and NNSA Award for Excellence. Dr. Peery earned his Ph.D. degree in nuclear engineering from Texas A&M University and joined Sandia National Laboratories as a member of the technical staff in 1990.

Benjamin Sawyer is co-founder of and gaming developer for Digitalmill, a games consulting firm in Portland, Maine. Mr. Sawyer has been involved in game development for over 10 years. He is also co-founder of the Serious Games Initiative (www.seriousgames.org) with the U.S. government's Woodrow Wilson International Center for Scholars in Washington, D.C. Founded in 2002, the initiative is one of the leading voices and organizers in the serious games field. In 2003, Mr. Sawyer started the first Serious Games Summit, a conference that now regularly attracts 300 to 500 attendees to discuss the latest best practices. In 2003 he co-founded the Games for Health Project, now the leading gathering of health care professionals, researchers, and game developers focused on creating health games and game simulations. As a developer, Mr. Sawyer has worked on over a dozen major serious game projects. He has been a designer, producer, advisor, or manager on projects for Cisco, the Defense Advanced Research Projects Agency, the Office of Naval Research, Leimandt Foundation, Cadbury, the U.S. Agency for International Development, Lockheed Martin, and several other Fortune 500 organizations. Digitalmill's work developing and advising serious game projects has been quite varied, including projects concerning command and control, education, advertising, training, and international development. Mr. Sawyer attended Baruch College in New York City but left short of his degree to pursue pressing professional offers.

Ethan Watrall is an assistant professor at Matrix: The Center for Humane Arts, Letters & Social Sciences Online; an assistant professor in the Department of Telecommunication, Information Studies, and Media; and an adjunct assistant professor in the Department of History at Michigan State University. In addition, he is a principal investigator in the Games for Entertainment & Learning Lab and co-founder of both the undergraduate specialization and game design development and the M.A. in serious game design at Michigan State University. Dr. Watrall's primary area of research is in the domain of cultural heritage informatics, specifically serious games for cultural heritage learning, outreach, and engagement. At Michigan State University, Dr. Watrall teaches in a wide variety of areas, including cultural heritage informatics, user-centered and user experience design, game design, serious game design, game studies, social history of popular culture, and ancient Egyptian social history and archaology.

Michael J. Zyda is director of the University of Southern California's GamePipe Laboratory and a professor of engineering practice in the Department of Computer Science. At USC he created the B.S. in computer science (games) and M.S. in computer science (game development) cross-disciplinary degree programs and doubled the incoming undergraduate enrollment of the computer science department. From 2000 to 2004 he was founding director of the MOVES Institute at the Naval Postgraduate School in Monterey, California, and a professor in the school's Department of Computer Science. From 1986 until the founding of the MOVES Institute, he was director of the NPSNET Research Group. Professor Zyda's research interests include computer graphics; large-scale, networked, three-dimensional virtual environments; agent-based simulation; modeling human and organizational behavior; interactive computer-generated stories; computer-generated characters; video production; entertainment/defense collaboration; modeling and simulation; and serious and entertainment games. He is a pioneer in the fields of computer graphics, networked virtual environments, modeling and simulation, and serious games. He holds a lifetime appointment as a national associate of the National Academies, an appointment made by the Council of the National Academy of Sciences in November 2003 in recognition of extraordinary service to the Academies.

Appendix B

Meetings and Speakers

MEETING 1
DECEMBER 3-4, 2008
THE KECK CENTER OF THE NATIONAL ACADEMIES
WASHINGTON, D.C.

DWO-4 Technology Forecasting
Steven Thompson, Chief, Technology Warning Division
Defense Intelligence Agency

Gaming Influences Culture
Mia Consalvo, Associate Professor, School of Telecommunications
Ohio University

Gaming: Past, Present, and Future
Gilman Louie, Partner
Alsop Louie Partners

Defense Modeling, Simulation, and Analysis: Meeting the Challenge
Neal Glassman, Senior Program Officer
Board on Mathematical Sciences and Their Applications
The National Academies

MEETING 2
JANUARY 12-13, 2009
THE BECKMAN CENTER OF THE NATIONAL ACADEMIES
IRVINE, CA

IBM Architecture
Laura Anderson, Program Director, Service Engineering
IBM Almaden Research Center

Serious Games Initiative
Ben Sawyer, President
Digitalmill, Inc.

Future of Modeling and Simulation
Drew Hamilton, President
Society for Modeling and Simulation International

Modeling and Simulation Capabilities
Thom Dunning, Director, Institute for Advanced Computing Applications and Technologies
University of Illinois at Urbana-Champaign

Trends Towards Convergence: Simulation, Games, Command and Control, Analysis, and the Historical Record
Jack Thorpe, Consultant

Learning Social Behavior and Language from Thousands
Jeff Orkin, Ph.D. Student
Massachusetts Institute of Technology

Games: The Virtual Frontier
Roger Smith, Chief Technical Officer, Program Executive Office for Simulation Training and Instrumentation
United States Army

Electronic Arts and the Birth of the Gaming Industry: A Conversation
Bing Gordon, Partner
Kleiner Perkins Caufield and Byers

Seriosity Games: A Conversation
Leighton Read, General Partner
Alloy Ventures, Inc.

APPENDIX B

MEETING 3
FEBRUARY 26-27, 2009
THE KECK CENTER OF THE NATIONAL ACADEMIES
WASHINGTON, D.C.

Future Capabilities Game 07 Results
Mort Rolleston, Onsite Strategic Planner for the U.S. Air Force AF/A8XC at SAIC
United States Air Force

Wargaming and the Future
Peter Perla, Director, Wargaming Strategic Linkage
Center for Naval Analyses

AI and Intelligent Tutoring
Brian Magerko, Assistant Professor
Georgia Institute of Technology

Gaming and Culture: Impact of the Internet
Timothy Mack, President
World Future Society

MMO's Presentation
Constance Steinkuehler, Associate Professor
University of Wisconsin, Madison

Social Networking
Cliff Lampe, Assistant Professor
Michigan State University

VC Presentation
Tim Chang, Principal
Norwest Ventures

Modeling and Simulation Vision
Russell Shilling, Scientific Advisor, Defense Center for Excellence
United States Navy

Policy Implications of Games: A Conversation
Dennis McBride, President
Potomac Institute for Policy Studies

3D Modeling, Simulation, Gaming, and Virtual World Technologies
Charlie Hargraves, 3D Immersive Virtual Environment Program Manager
Lockheed Martin

**MEETING 4
APRIL 6-7, 2009
THE JONSSON CENTER OF THE NATIONAL ACADEMIES
WOODS HOLE, MA**

Writing Meeting

Appendix C

Committee Methodology

KEY FEATURES OF THE METHODOLOGY

A robust methodology for technical inquiry should have four key features. First, to be accepted, it must be presented in a lexicon and structure appropriate for the user's culture—in this case, for the culture in which the Defense Intelligence Agency (DIA) Technology Warning Division (the sponsor of this study) operates. Any communication of findings, conclusions, or recommendations offered by the committee must be expressed accordingly. The division makes use of weather-forecasting terminology (Futures, Watch, Warning, Alert)[1] in the issuance of technology assessments, making the overall warning message regarding all products readily interpretable by any reader. The committee adopted and adapted the DIA's vocabulary to characterize the relative status—and recommended action—for each technology.

The second key feature is that, to be relevant, the study methodology must be tied in a fundamental way to top-level Department of Defense (DOD) strategies. For example, the committee reviewed Joint Vision 2020 (JCS, 2000) to validate its selection of the technology topics addressed in this report. In future studies, to facilitate integration into the larger body of intelligence materials, the committee proposes that technology selections be derived through a more disciplined, RED team[2] review of top-level strategy documents (e.g., Joint Vision 2020) with an eye to identifying technologies that could be used to deny a BLUE[3] capability deemed critical to U.S. military success.

The third key feature is that, to maintain focus and ensure timeliness, the study methodology must yield assessments built on a solid understanding of the technical feasibility of potential technology-based threats. This requirement leads to a capability-based approach for investigating and categorizing candidate technologies. Furthermore, the technical peer review process to which all NRC reports are subjected provides additional assurance of the technical quality of committee assessments.

NOTE: The methodology for technology warning described in this appendix is reprinted from *Avoiding Surprise in an Era of Global Technology Advances* (Washington, D.C.: The National Academies Press, 2005), pp. 20-27.

[1] The definitions used by the DIA for these terms are as follows: Futures—Create a technology roadmap and forecast; identify potential observables to aid in the tracking of technological advances. Technology Watch—Monitor global communications and publications for breakthroughs and integrations. Technology Warning—Positive observables indicate that a prototype has been achieved. Technology Alert—An adversary has been identified and operational capability is known to exist.

[2] "RED" is used in this report to denote the adversary or an adversarial perspective (e.g., "RED team").

[3] "BLUE" is used in this report to denote U.S. military forces.

Lastly, to be enduring, the methodology should accommodate evolving realities of science and technology (S&T) leadership, driven by the synergistic trends of globalization and commercialization described in Chapter 1. Traditionally, the United States has assumed that it leads the world in S&T. This perspective leads the technology warning community to look for indications that external actors are trying to "catch up," or to exploit known technologies in new ways. Projected trends suggest that it should no longer automatically be assumed that the United States will lead technological advances in all relevant technologies. This reality imposes a new burden on the technology warning community, generating the need for it to search in different places and in different ways for the information needed to warn against technology surprise.

FOUNDATION OF THE METHODOLOGY

The committee believes that the Technology Warning Division can most effectively prioritize its limited resources by utilizing a capabilities-based approach with respect to assessing technologies. The landscape of potentially important emerging technologies is both vast and diverse. Ideally, the division should assess whether a given technology has the potential to pose a viable threat prior to commissioning in-depth analyses. Since the division is keenly interested in when specific technologies may mature to the point that they pose a threat to U.S. forces, a functional decomposition from an adversarial, or RED, perspective is most useful. The methodology defined by the committee begins with the following focus question: *What capabilities do the United States have that, if threatened, impact U.S. military preeminence?*

In general, U.S. capabilities could be threatened either through direct denial of or disruption of BLUE capabilities or via RED capabilities that negate or significantly diminish the value of BLUE capabilities (e.g., improvised explosive devices (IEDs) being employed by insurgent forces in Iraq).

Joint Vision 2020 was used to define the basic framework for U.S. military capabilities deemed vital to sustained success (JCS, 2000). The overarching focus of this vision is Full Spectrum Dominance—achieved through the interdependent application of four operational concepts (Dominant Maneuver, Precision Engagement, Focused Logistics, and Full Dimensional Protection) and enabled through Information Superiority, as illustrated in Figure [C]-1 (JCS, 2000).

The committee selected the four operational concepts, together with Information Superiority, as the foundation for its assessment methodology. Joint Vision 2020 provides the definitions presented in Box [C]-1.

The committee also noted the importance of technology warning with respect to the "Innovation" component of Joint Vision 2020 shown in Box [C]-2, since "leaders must assess the efficacy of new ideas, the potential drawbacks to new concepts, the capabilities of potential adversaries, the costs versus benefits of new technologies, and the organizational implications of new capabilities" (JCS, 2000).

From this foundation the committee then identifies specific capabilities in accordance with the previously defined focus question—*What capabilities does the United States have that, if threatened, impact U.S. military preeminence?*

While the U.S. military has devoted significant time to the definition of vital capabilities in alignment with Joint Vision 2020, the committee made no effort in this first report to synchronize its derivations or definitions, or to provide a complete decomposition of the operational concepts and enablers into their underlying capabilities. Rather, committee members selected a few evolving technologies and assessed the potential for those technologies to threaten important U.S. capabilities. Given that the committee's proposed basic methodology is adopted, future studies will analyze more comprehensively the threats to a taxonomy of U.S. military capabilities that derives from the operational concepts envisioned by Joint Vision 2020. The basic methodology developed by the committee is summarized in Box [C]-1 and is described in greater detail in subsequent sections.

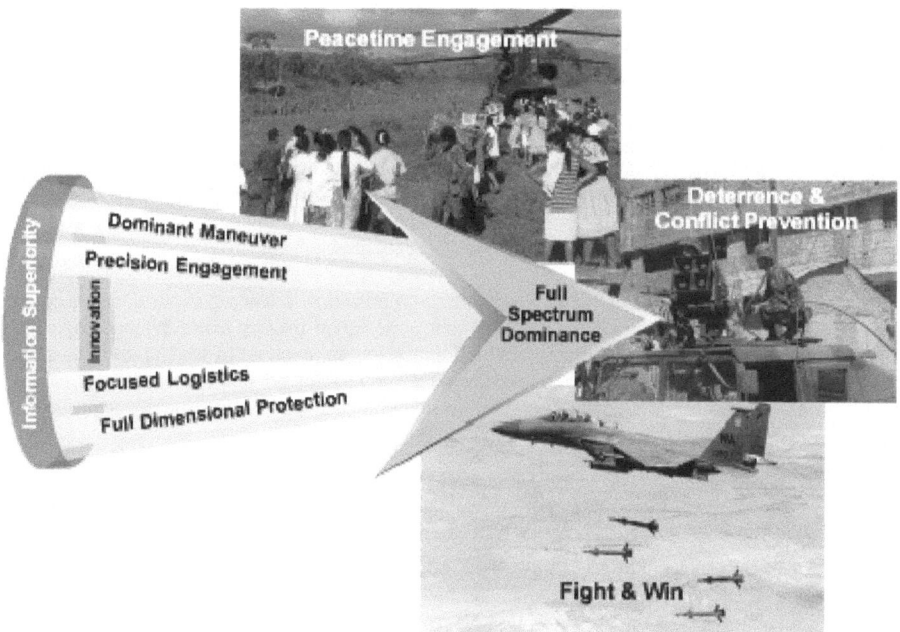

FIGURE [C]-1 Concepts constituting the basic framework for U.S. military capability as defined by Joint Vision 2020. (See Box [C]-1.) SOURCE: JCS (2000).

IDENTIFY

The next step of the proposed assessment methodology is performed from the RED perspective. The central question here is as follows: *What are the evolving technologies that, in the hands of U.S. adversaries, might be used to threaten an important U.S. military capability?* A corollary question is, What technologies, if rapidly exploited by the U.S. military, are likely to yield sustained technological superiority? However, this issue was addressed only peripherally, given the division's focus on technology warning.

Having identified a technology of potential interest, the next challenge becomes the derivation of "indicators" or "observables" that may suggest adversarial adoption or exploitation of that technology. Although targeted intelligence-collection methods remain important, in this report the committee focuses on observables that may be derived from open source analysis—leveraging the effects of the information revolution and acknowledging that the twin forces of globalization and commercialization provide new sources of relevant information. At the same time, however, the committee recognizes the difficulty of discerning when technological advances portend emerging threats rather than societal benefits.

A sample chart—Chart [C]-1—exemplifies how each technology is assessed.

ASSESS

The committee's assessment methodology involves characterization of a technology in terms of three variables: Accessibility, Maturity, and Consequence. Priorities for more detailed analyses may derive from any individual variable or any combination of the three.

> **BOX [C]-1**
> **Relevant Definitions from Joint Vision 2020 Serving as Foundation for Assessment Methodology**
>
> **Information Superiority** is the capability to collect, process, and disseminate an uninterrupted flow of information while exploiting or denying an adversary's ability to do the same. Information superiority is achieved in a noncombat situation or one in which there are no clearly defined adversaries when friendly forces have the information necessary to achieve operational objectives.
>
> **Dominant Maneuver** is the ability of joint forces to gain positional advantage with decisive speed and overwhelming operational tempo in the achievement of assigned military tasks. Widely dispersed joint air, land, sea, amphibious, special operations and space forces, capable of scaling and massing force or forces and the effects of fires as required for either combat or noncombat operations, will secure advantage across the range of military operations through the application of information, deception, engagement, mobility and counter-mobility capabilities.
>
> **Focused Logistics** is the ability to provide the joint force the right personnel, equipment, and supplies in the right place, at the right time, and in the right quantity, across the full range of military operations. This will be made possible through a real-time, web-based information system providing total asset visibility as part of a common relevant operational picture, effectively linking the operator and logistician across Services and support agencies.
>
> **Precision Engagement** is the ability of joint forces to locate, surveil, discern, and track objectives or targets; select, organize, and use the correct systems; generate desired effects; assess results; and re-engage with decisive speed and overwhelming operational tempo as required, throughout the full range of military operations.
>
> **Full Dimensional Protection** is the ability of the joint force to protect its personnel and other assets required to decisively execute assigned tasks. Full dimensional protection is achieved through the tailored selection and application of multilayered active and passive measures, within the domains of air, land, sea, space, and information across the range of military operations with an acceptable level of risk.
>
> SOURCE: JCS (2000).

Accessibility

The Accessibility variable focuses on the question *How difficult would it be for an adversary to exploit the technology?* It addresses the ability of an adversary to gain access to and exploit a given technology. This assessment is divided into three levels:

- *Level 1*. The technology is available through the Internet, being a commercial off-the-shelf item; low sophistication is required to exploit it.
- *Level 2*. The technology would require a small investment (hundreds to a few hundred thousand dollars) in facilities and/or expertise.
- *Level 3*. The technology would require a major investment (millions to billions of dollars) in facilities and/or expertise.

In general, Level 1 technologies are those driven by the global commercial technology environment; they are available for exploitation by a diverse range of potential adversaries. Level 3 technologies, by contrast, are typically accessible only to state-based actors. The indicators likely to be of value in determining an adversary's actual access to a given technology vary by level as well as by the type of technology.

BOX [C]-2
Proposed Methodology for Technology Warning

Foundation Joint Vision 2020[a] Operational Concepts and Information Superiority

- **Focus** *What capabilities does the United States have that, if threatened, impact U.S. military preeminence?*

- **Identify** *What are the evolving technologies that, in the hands of U.S. adversaries, might be used to threaten an important U.S. military capability?*

  What are the observables that may indicate adversarial adoption or exploitation of such technologies?

- **Assess** *Accessibility: How difficult would it be for an adversary to exploit the technology?*

  Maturity: How much is known about an adversary's intentions to exploit the technology?

  Consequence: What is the impact on U.S. military capability should the technology be employed by an adversary?

- **Prioritize** *Identify: What are the relative resources to be applied to each emerging technology to support the technology warning process?*

- **Task** *Establish and assign intelligence-information-collection requirements.*

[a]SOURCE: JCS (2000).

CHART [C]-1 Example of Technology Assessment Chart

Technology		Observables
Brief description of technology		Brief description of observables
Accessibility	Maturity	Consequence
Level 1, 2, or 3	Technology Futures Technology Watch Technology Warning Technology Alert	Short characterization

Maturity

The Maturity variable focuses on the question *How much is known about an adversary's intentions to exploit the technology?* It integrates what is known about an adversary's actions, together with an evaluation of the state of play with respect to the technology of interest. At the highest level, called Technology Alert, an adversary has been identified and an operational capability has been observed. At the lowest level, Technology Futures, the potential for a technology-based threat has been identified, but no positive indicators have been observed. The Maturity assessment is divided into four categories: the first two (the lower levels) suggest further actions for the technology warning community; the other two indicate the need for immediate attention by military leadership:

- *Futures.* Create a technology roadmap and forecast; identify potential observables to aid in the tracking of technological advances.
- *Technology Watch.* Monitor (global) communications and publications for breakthroughs and integrations.
- *Technology Warning.* Positive observables indicate that a prototype has been achieved.
- *Technology Alert.* An adversary has been identified and operational capability is known to exist.

Given the potential for disruptive advances through technological breakthroughs or innovative integration, as well as the difficulty of identifying and tracking meaningful indicators, any particular technology is unlikely to progress sequentially through the various categories of Maturity listed above.

As indicated at the beginning of this chapter, the committee adopted and adapted the DIA's terminology in defining these categories. The definitions are likely to evolve as the process matures. The committee sees significant value in this basic approach, however, since it divorces the challenge of technology warning from the discrete time lines associated with "prediction," which are almost invariably inaccurate.

Consequence

Characterization of a technology in terms of the Consequence variable involves addressing the question *What is the impact on military capability should the technology be employed by an adversary?* It involves assessing the impact of the postulated RED technology on the capability of BLUE forces. This impact can range from denial or negation of a critical capability to the less-consequential level of annoyance or nuisance. A corollary assessment may be made as to the locus of impact—that is, whether the technology affects a single person, as in the case of an assassination, or creates a circumstance of mass casualty and attendant mass chaos.

PRIORITIZE

The objective of the prioritization step of the methodology is to respond to the question *What are the relative resources to be applied to each emerging technology to support the technology warning process?* This step is intended to harmonize the distinct nodes of observed capability, demonstrated intent, resources available, and the inherent cost of inaction. Prioritization is key to the technology warning methodology, since the Technology Warning Division lacks the resources to fully analyze every conceivable evolving technology. It is equally important to recognize that prioritization is an integral part of each methodology parameter. The prioritization of individual parameters is based on the levels of change detection and potential impact. By prioritizing the parameter, the division can focus subsequent analyses over a smaller subset of an assigned change detection domain. Priority assignment is essential to enable the focusing of more sophisticated information-gathering tools and analytic techniques on the areas of highest potential concern.

The prioritization methodology lends itself to any number of commercially available tools and techniques designed for assistance in establishing and maintaining a logical and consistent focus as well as the flexibility to react to the dynamics of technology change and country-of-interest variability. During the prioritization process, it will be important to establish measures of performance to allow critical analysis as well as change management in order to improve the overall process. The end result of the prioritization process is to provide for actionable awareness with which to influence analysis and tasking, the last of the methodology parameters.

The committee envisions that prioritization would be accomplished in close consultation with the technology warning community. It made no attempt to further develop the prioritization process in this report.

TASK

The Technology Warning Division will inevitably have unmet needs for additional information and/or intelligence relating to the prioritized list of evolving technologies. Although some needs may be met through division-chartered research, others will require the assistance of the broader intelligence community.

The task step—*Establish and assign intelligence-information-collection requirements"*—involves the dissemination of collection requirements to other IC components and subordinate agencies. Such requirements must provide sufficient specificity to enable interpretation by collectors who are not necessarily literate in the specific technology. The requirements may include general instructions for accomplishing the mission. It is envisioned that some of the observables postulated in the Identify step of the methodology will provide a useful basis for such tasking.

The results from collection efforts will be integrated back into the assessment step in order to refine, reprocess, and update the division's understanding of a given technology. This analysis may stimulate the issuance of a new report to the division's customers to inform them of changes in the assessed maturity of that technology.

USING THE METHODOLOGY IN THIS REPORT

To test the robustness of the proposed technology warning methodology, the committee applied it in order to assess four key areas in this initial report. It should be noted that this initial exercise was necessarily circumscribed by the domain expertise represented in the committee members and by the shortness of time for broader outreach to the technical community at large. Furthermore, since the methodology emerged in parallel with the committee's technology assessments, the approaches taken were not entirely consistent.

The foundation provided by Joint Vision 2020 and augmented by the military and professional backgrounds of committee members was used to select the following four key capabilities to assess:

- Information superiority (Chapter 3),
- Air superiority (Chapter 4),
- Discrimination between friends/foes/neutrals (Chapter 5), and
- Battle readiness and communications superiority (Chapter 6).

Chapters 3 through 6 each address the "Identify" activity with examples of evolving technologies that may threaten the capability and potential indicators that such technology development is underway. The "Assess" activity then examines opportunity and motivation for adversarial technology development and/or employment, posits change detection relative to the indicators, and assesses likely impact. Preliminary characterizations of accessibility, maturity, and consequence are provided for most evolving technologies, although the level of specificity is variable.

Subsequent steps (i.e., "Prioritize" and "Task") of the proposed methodology require customer inputs and actions and are left to future study efforts.

REFERENCE

JCS (Joint Chiefs of Staff). 2000. Joint Vision 2020. Director for Strategic Plans and Policy, J5; Strategy Division. U.S. Government Printing Office, Washington, D.C. June.

Appendix D

Key Recommendations from Previous Studies

The committee referenced several previous National Research Council studies related to modeling and simulation. Table D-1 lists the studies, which are summarized in this appendix.

TABLE D-1 Previous Modeling and Simulation Studies

No.	Study Name	Prepared by	Publication Year
1	Modeling and Simulation: Linking Entertainment and Defense	Committee on Modeling and Simulation: Opportunities for Collaboration Between the Defense and Entertainment Research Communities	1997
2	Defense Modeling, Simulation, and Analysis: Meeting the Challenge	Committee on Modeling and Simulation for Defense Transformation	2006
3	Behavioral Modeling and Simulation: From Individuals to Societies	Committee on Organizational Modeling: From Individuals to Societies	2008

APPENDIX D

SUMMARY 1
MODELING AND SIMULATION:
LINKING ENTERTAINMENT AND DEFENSE[1]

Background

In recent years, modeling and simulation technology has become increasingly important to both the entertainment industry and the U.S. Department of Defense (DOD). In the entertainment industry, such technology lies at the heart of video games, theme park attractions, entertainment centers, and special effects for film production. For DOD, modeling and simulation technology provides a low-cost means of conducting joint training exercises, evaluating new doctrine and tactics, and studying the effectiveness of new weapons systems. Both the entertainment industry and DOD are aggressively pursuing development of distributed simulation systems that can support Internet-based games and large-scale training exercises. These common interests suggest that the entertainment industry and DOD may be able to more efficiently achieve their individual goals by working together to advance the technology base for modeling and simulation. Such cooperation could take many forms, including collaborative research and development projects, sharing research results, or coordinating ongoing research programs to avoid unnecessary duplication of effort. (p. 1)

In October 1996 the National Research Council's Computer Science and Telecommunications Board held a workshop that brought together members of the entertainment and defense industries to discuss common research interests in modeling and simulation. These discussions illuminated possible areas for cooperation and also brought up for consideration possible cultural and logistical obstacles to collaborative success.

Conclusions

The following research areas were identified as possible areas of collaboration between the defense and entertainment industries:

- **Technology for immersion:** Immersive virtual environments rely on a cluster of technologies that include graphics software for creating complex visual environments, and technologies that track participants' physical orientation and gaze, facilitate the creation of realistic virtual terrain, and provide users with sensory stimuli (sounds, smells, vibration, etc.).
- **Networked simulation:** DoD and the entertainment industry both seek to create a network infrastructure capable of handling large-scale networks of users. This will require developing higher bandwidth networks to handle the flow of large amounts of data, reducing bandwidth requirements using multicasting and area-of-interest managers, and minimizing signal latency.
- **Standards for interoperability:** In order to maximize the usefulness of simulations, both groups should be able to work with other programs in a meaningful and coherent fashion. Interoperability requires common network software architecture with standard protocols that allow interaction between simulators and facilitate the construction of large simulations from existing subsystems. The development of a virtual reality transfer protocol is also crucial to facilitate large-scale networking of distributed virtual environments.
- **Computer-generated characters:** One of the greatest challenges in creating a useful simulation,

---

[1]The first paragraph of this summary is excerpted from *Modeling and Simulation: Linking Entertainment and Defense* (Washington, DC: National Academy Press, 1997).

the development of realistic computer-generated characters presents a number of research and development challenges. First is adaptability, or the ability of a virtual character to learn new behaviors in response to changes in environment and input. The next challenge is developing computer-generated characters that look, move, and express emotion like real humans.
- **Tools for creating simulated environments:** Better hardware and software tools are needed for creating realistic virtual environments. Specifically, this involves the development of better tools for the construction, manipulation, and composition of large databases of information and facilitation of composite visual image creation. Designers would also like to see input devices more sophisticated than the computer mouse and keyboard for use in designing environments.

In addition, the committee identified three areas that should be given special attention during the planning and execution of any sustained collaborative effort:

- **Information sharing and technology transfer:** The two communities that participated in this workshop are known for having cultures that discourage information sharing and transfer little technology. Although there is little precedent, mutually beneficial transfers of information and technology can occur in the form of formal collaborative arrangements between entertainment companies and DoD or by encouraging professionals in DoD to attend entertainment industry conferences and vice-versa.
- **Human resources:** There is a shortage of talented people with the broad range of skills required to create successful simulations, and few university programs teach this skill set. To remedy the situation, existing funding mechanisms could be used to enhance educational programs that combine technical and artistic studies.
- **Preserving the research base:** Ensuring an adequate supply of new ideas and technologies for modeling and simulation requires continuous support of basic research. Growing demands for accountability in government funding have had the effect of restricting the amount and scope of research being conducted. Industrial contributions to such research have also started to wane. These issues must be addressed to ensure the continued viability of the technological base for modeling and simulation.

Strong commonalities exist between defense and entertainment applications of modeling and simulation and the technologies needed to support them. Aligning the research agendas of these two communities to allow greater coordination of research developments, information sharing, and collaborative research could provide an opportunity to more rapidly achieve the goals of both industries. Although linking these two communities presents a significant challenge, sustained commitment from both sides will go a long way in ensuring successful collaboration. This workshop was a first step in examining the costs and benefits of such a commitment.

SUMMARY 2
DEFENSE MODELING, SIMULATION, AND ANALYSIS: MEETING THE CHALLENGE

Background

Since World War I, modeling, simulation, and analysis (MS&A) has been an important tool of the U.S. Department of Defense (DoD). The current legacy systems of MS&A, however, are insufficient to deal with new adversarial challenges. In addition to planning for conflict with the armed forces of other nations, the U.S. military has been called on to confront insurgents and terrorists, often in urban environments where civilian noncombatants are prevalent. Because of these changes, DoD now regularly includes diplomatic, intelligence, military, and economic tactics in decision making. Coinciding with this change in environment is the emergence of a new generation of interconnected, interdependent military systems, including unmanned weapons, and the rise of network centered warfare, all of which increase the potential for unforeseen consequences in combat. As military technology becomes more advanced and network oriented, and as noncombat activities become a larger part of the nation's defense strategy, there is a critical need for comprehensive, state-of-the-art MS&A capabilities to aid in understanding the behavior of both hardware/software networks and social networks.

A report was commissioned by the Modeling and Simulation Coordination Office (formerly known as the Defense Modeling and Simulation Office) to evaluate current DoD MS&A capabilities, to identify the research areas with the greatest potential to benefit future MS&A systems, and to recommend ways to communicate the risks and benefits of current and potential MS&A to decision makers and other nonexperts.

Findings and Recommendations

The committee found that the current MS&A systems in operation are inadequate. They employ scripted nonadaptive scenarios and do not incorporate the body of existing knowledge on phenomena related to terrorist and insurgent networks. The committee recommended that the development of a body of robust, flexible, and adaptive MS&A strategies be given priority by DoD. Also, it recommended that the following objectives guide DoD's MS&A efforts:

- **Focus on network-centric operations:** DoD should establish a comprehensive and systematic approach for developing MS&A capabilities to represent network-centric operations by fostering collaboration between the parties developing MS&A systems, and further developing existing approaches to network-centric operation while incorporating new approaches and mathematical models.
- **Maintain a broad research portfolio:** Significant research should be devoted to a wide variety of data collection methods, models, and games in order to meet the diversity of challenges faced by the DoD MS&A community. These should include investigation into social behavioral networks, multiagent systems, and game-based training.
- **Create a dedicated workforce:** DoD should assign an organization the responsibility of developing and supporting research for use in combat and noncombat modeling. It would also be responsible for ensuring that students and practitioners of MS&A are exposed to all core disciplines involved in the creation of MS&A systems and that students are able to use their knowledge of MS&A to address problems of importance to the consumers of their product.
- **Minimize uncertainty:** DoD should seek better methods to characterize, quantify, and manage the uncertainty inherent in all modeling and simulation, including inputs, modeling assumptions, parameters, and options.
- **Be aware of cognitive styles:** MS&A practitioners should strive to understand the cognitive styles of decision makers, as well as their interaction with different styles of MS&A. This would both inform the creation of the MS&A applications and aid in communicating the usefulness of MS&A to critical nonexpert audiences.

In conclusion, the committee emphasized the need for information sharing and collaboration in the DoD MS&A community. DoD's current MS&A apparatus is widely distributed across many offices and programs, and the committee strongly suggested that it be consolidated into a single dedicated office. This office would be a natural adjunct to the Modeling and Simulation Coordination Office and would create an intellectual common ground for the military MS&A community necessary for the achievement of all previously listed recommendations.

SUMMARY 3
BEHAVIORAL MODELING AND SIMULATION: FROM INDIVIDUALS TO SOCIETIES

Background

Instead of fighting nation-states with conventional weapons, today's U.S. military is increasingly called on to confront insurgents and terrorist networks in highly populated areas, where the attitudes and behaviors of civilian noncombatants are greatly affected by military actions. In order to achieve victory, warfighters must be able to selectively apply combat tactics, administer humanitarian aid, and peacefully mediate conflicts. The Air Force and other military services have a growing need for individual, organizational, and societal (IOS) behavioral models in order to inform the development of doctrine, strategies, and tactics for dealing with adversaries, for use in training and mission rehearsal, and in the analysis of current political and military situations.

The Committee on Organizational Modeling was conceived at the request of the U.S. Air Force to evaluate the current state of the art in the IOS modeling research areas best suited to military applications. In addition, the committee was asked to identify weaknesses in the current DoD computational models of behavior. The committee members would then suggest ways in which these weaknesses might be addressed through the use of other systems and through better collaboration with the social science community, and recommend a roadmap for the research and development of an improved modeling system for the near, mid, and far term.

Conclusions

Human behavioral modeling is a complex science with roots in several disciplines. Rather than having a single correct model, the field encompasses a collection of potentially useful approaches. The committee believes a multilevel modeling approach combining several existing models linked together is the most promising for addressing the complex challenges faced by DoD.

The committee identified five potential problem areas to be considered when developing IOS models:

- Problems in modeling strategy can arise due to unrealistic expectations about how faithful a model can be to the real world.
- Standards of verification, validation, and accreditation designed for use with models of physical systems cannot always be usefully applied to behavioral and social models.
- The scope of the human behavior domain (social, organizational, cultural, or individual) must be appropriately matched to the phenomena being modeled.

- Uncertainty reduction is less plausible in models of human behavior than in models of physical phenomena. Structures and processes can undergo rapid change, making adaptation a key feature of a model and its users.
- Combined models using different approaches and distinct behavioral domains must be linked carefully to ensure effectiveness.

Recommendations

The committee recommended that the sponsor fund cross-disciplinary research programs focused on representative challenges in IOS modeling, using common datasets. In addition, research efforts should be expanded in six specific areas to advance modeling capabilities in the long term:

- Theories for basic social behaviors need to be more fully developed to provide the basis for models of large-scale social patterns.
- Basic research is needed to address challenges particular to IOS modeling, such as how models can capture human adaptation over time through learning, rational and irrational behavior, and the effects an individual's competing allegiances have on decision making.
- Data collection methods are in need of development and refinement.
- More investigation into federated models is needed in order to ascertain which components of a model should be linked and which should be encapsulated, how different classes of models should be linked to one another, and so forth.
- New, clear guidelines should be established for validation standards based on the IOS model's goals and purpose. The standard of "validation for action" would be distinct from the standard currently used for models in the physical sciences.
- Model building should be facilitated through the construction and ongoing maintenance of national Web-accessible data repositories and an online catalog of general approaches, models, simulations, and tools for developers.

Finally, the committee recommended that researchers, developers, and government representatives meet regularly at multidisciplinary workshops and conferences to compare and exchange results. This would facilitate collaboration and growth in the currently fragmented field of IOS modeling and accelerate progress on some of the above listed challenges.

Appendix E

An Overview of Digital Games

As the intelligence community and others think about positioning themselves to utilize and understand games, they must first understand the range and language of the game types, players, and gaming experiences they might pursue. Beyond recognizing these available game play tools, communities must work to understand the range of outputs of gaming and potential applications beyond the entertainment world.

This appendix gives an in-depth overview of two important aspects of games. The first segment of the appendix looks at the social and cultural outputs of games, covering specifically the range of genres, platforms, and player types of games. The latter half examines distinct segments of the games industry and analyzes the relevance of its various business models. This growth emphasizes the pervasiveness and still unrealized potential of the games industry.

GAME TYPES

Games yield from a number of distinct genres and on a variety of platforms, all with specific conventions. These genres types are often attached to specific design patterns, the root mechanics that define game play. The committee provides example descriptions of several genres to illustrate the scope and diversity of game genres, not only from a technological and design perspective but from a usage and social perspective as well. The following categorizations dive into the genres and medium for play as well as player constructs.

Genres

Much like other forms of entertainment media, digital games are often categorized by genre. The difference, in the case of digital games, is that game play interactions rather than visual or narrative differences are used as the prime organizing principle. As the digital game medium has become more mature, the lines between genres have become increasingly blurred—a phenomenon similar to that which has occurred for television and film. As a result, the descriptions of genres provided—listed below—should be taken as a general framework.

- Desktop simulation games
- War games
- Puzzles
- Adventure
- Role-playing games
- Action
- First-person shooters
- Strategy
- "God" games and other simulation games
- Alternate-reality games
- Serious games

Desktop Simulation Games

Since the dawn of computer games there have always been games best described more as "simulations" in that they tended to be purposely built scaled software models of real-world systems or devices. While simulation may fit the definition of many computer games in the world of computer and video games, a simulation game is more often than not a nonfiction vehicle simulation usually of military persuasion. That said, the world of computer game simulations is increasingly filled with games that use the core design pattern of desktop vehicle simulation but that feature fictional vehicles such as X-Wing Fighters and futuristic tanks, among others. Some entire genres of games are quite simulative but from a genre distinction standpoint are grouped under different banners, such as sport games (see *Madden Football*), racing games (see *Need for Speed* or *DIRT*), and many strategy titles (e.g., *SimCity*, *Rise of Nations*).

In the simulation genre as defined mostly by vehicle simulation, the computer games world has had a fair amount of activity. Microsoft's *Flight Simulator* has been a widely respected simulation product that even saw some derivative usage for large nonentertainment simulation usage (as Microsoft ESP) before the entire product and its derivatives were recently shut down in a reorganization of Microsoft's games business. Throughout the years the games industry has produced strong desktop simulations of F-15/16/18 fighters and many other aircraft, various tanks, naval ships (PHM Pegasus, Strike Fleet), space vehicles, civilian cars, and much more. Originally these simulations were fairly scaled compared to counterparts on workstation systems or higher end computing systems only available to government or large corporate customers. However, paralleling the rise of desktop computing as it supplanted these systems, the core computer games simulation genre has itself matured considerably. There is little difference in graphics and some game play modeling with today's desktop simulations of vehicles and those played on non-personal computer (PC) systems. The architectures are now fairly unified and the software is at times almost the same, the only distinction being some level of realism and graphical response made possible by parallel computing platforms running the same software at a higher resolution and with more modeling or graphical effects turned on.

Like the war games genre, the "pure simulation" genre in computer games is a smaller percentage of the overall games market, as other genres like sports, action games, and so forth, have arisen. This has placed the simulation genre in a more niche realm, which while smaller in some market-size respects is heavily populated by very devout user bases. Over time these user bases are increasingly supported by smaller dedicated developers who service the genre and keep advancing it. Where there are some breakouts from the more entrenched smaller market, they tend to be in very specific areas such as battlespace games (see *Operation Flashpoint*, *VBS*, etc.) that combine vehicle simulation with traditional first-person soldier battle or sister genres like sports and racing.

Overall the core genre and its cousins do have a clear dynamic that realism is a critical selling point. This often starts with graphical fidelity but quickly also includes fidelity of various interfaces and critical modeling such as weather, flight dynamics, vehicle capabilities, and more. It is this selling point of realism that drives the creators and the core fan base in the genre. As one of the first genres to really embrace multiplayer play, there is also increasingly little need for computer artifical intelligence (AI) and any paradoxes of realism that a game developer can create in the simulation genre. Thus, realism very much is about how realistically rendered the simulation is leaving the simulation of the opponents to other humans versus a machine that must ultimately lose to a less skilled player. What desktop game simulations represent to the larger field of simulation is the continued evolution of constrained but highly capable simulations people may learn from—flight simulator offers some level of help to those learning to fly and other simulations may provide similar if still small coefficients to real-world training and operation of those vehicles being simulated.

War Games

In the world of video games, war games started out essentially as computer-powered tabletop war games. Many of the original designers were essentially those that moved over from venerable tabletop game creators like Avalon Hill to computer-based games. The main advantage of the computer was twofold. First, the game could rapidly compute outcomes of units engaged and, perhaps more importantly, the computer could use basic approaches to AI to create single-player experiences. This allowed war games and general strategy game cousins to flourish beyond their board game pedigree because now you could play them by yourself, dispensing with the need for correspondence (no e-mail in those days) or the presence of another capable player. Companies such as SSI (Strategic Simulations, Inc.) and even board maker Avalon Hill published dozens of war games during the early to late 1980s.

The basic computer war game design pattern is essentially one where opposing forces deploy units and move them on a game board and, whenever two units attempt to occupy the same piece of land (usually a spot on a $X \times X$ grid), there is a battle where one or more factors are used to calculate an outcome. Sides each take turns moving their units, and the usual factors for each unit are movement speed (how many units it can move in one turn), strength (i.e., firepower), defensive capability (how likely they can withstand attack), and often morale, which acts as some level of positive or negative force multiplier. Most computer war games have been historical in nature, but as titles evolved, there have been a larger number of fictional or even generalized titles.

As programmers and designers began thinking beyond the first advantages of computerized board games, they started to create improved aspects of play for war games played on computers. Chris Crawford, at Atari, created *Eastern Front 1941,* in which a player plotted paths for all units at once and then watched as the computer played out the resulting skirmishes. Other designers added more factors to the general calculus of war games as well as ever more elaborate editors for others to use. AI has also improved but an important aspect of AI for war games as well as many other computer games is that the AI is intended as an obstacle that can be overcome by the player and so the approach of creating an AI that is "Deep Blue" in nature is often not the goal of many war game designers. So AI has advanced to create more realistic approaches that challenge the player but not so the player is overwhelmed. Indeed, when Origin Systems created a computer version of Steve Jackson Games' *Ogre* it had to dumb down the AI because the computer was too good and thus the game was deemed by play testers as unplayable.

Despite better graphics, increased strategy, and better marketing, computer war games have remained a smaller niche than many other genres of games. Only in two respects have war games moved beyond their niche today. First, they contribute generally to the large parent genre of strategy games (and even some role-playing games) that contain lots of general principles and interface elements of the first generations of war games. This includes games such as Ubisoft's *Heroes of Might & Magic*, *Civilization*, Nintendo's *Advance Wars*, and Square-Enix's *Final Fantasy Tactics*.

Second, where computer war games have most evolved to break out of their niche status is as "real-time strategy" games, or RTS games as they are more readily called. The idea of real-time strategy games is to essentially do away with the turn-based approach that has and still dominates many war games today. Their approach is to reward those players who can muster strategy and resources and physically command them faster against an enemy. Any unit can move at any time based on a commander's ability to tell them what to do. The RTS has evolved out of war games into its own recognized genre of video games.

The first RTS game was arguably Herzog's Zwei on the Sega Genesis in the late 1980s. This was followed by the Westwood Games (now part of Electronic Arts) title *Dune2*, which then beget one of the stalwart RTS brands, *Command & Conquer*. This top title was later joined by Blizzard's *Warcraft* and *Starcraft* series and Microsoft's *Age of Empires* series. Overall the success of the RTS has further pigeonholed the more traditional computer war game as the RTS has sold well enough to get far better budgets, graphics, and marketing, thereby reinforcing its success as a game genre.

The irony of the RTS, however, is that it is a perverted sense of real time. No version of warfare past small squad-based encounters actually unfolds in the seconds and compressed time of any RTS. Traditional war at the theater level plays out in hours/days/weeks versus seconds and minutes. A true "real-time" war game would mimic the 1:1 nature of modern conflict, which ironically would bore most video gamers used to instant feedback, quick resolution of conflict, and fast iterative play involving trial and error. While it may demand a bit more cognitively of players, there remains a question of whether the RTS style of war games offers anything important to more methodological war games played out in the military.

Today the core war game genre of detailed historical or fictional conflicts played out over large theaters of war is essentially still a purview of enthusiasts, history buffs, and fans of the genre known collectively as "grognards" (http://en.wikipedia.org/wiki/Grognard), which in the 1970s through today essentially is a term for those who love the war games genre.

The importance of computer war games as a genre is twofold. First, it is the historical nature of how war games link more advanced styles of game design from pen-and-paper origins to that of central-processing-unit-based gaming and how the thread of war gaming runs through many evolutionary stages of game development in the modern video game era. Second, and perhaps useful today, is the fact that war games have at times seen some levels of advancement on the video games side more so than in other areas where they are used. Even in their reduced market size, there has been strong continued evolution of the genre, including games like *Harpoon, Panzer General, Empire, Patton vs. Rommel,* the *Total War Series*, and the work of stalwarts such as *Slitherine Simulations* (http://www.slitherine.com), which have continued to seek ways where core game play is enhanced through better simulations, models, graphics, and interface. What this might tell those for whom war games are important is that despite their overall reduced commercial capacity it is still that continued consumer entertainment-oriented progression that offers large utility to projects utilizing war gaming for more specific nonentertainment needs.

Puzzles

As the name implies, puzzle games generally revolve around a player solving a puzzle. The types of puzzles can vary wildly, and they require players to engage in tests of logic, strategy, pattern recognition, sequence solving, spatial awareness, and word completion (Rollings and Adams, 2003). Perhaps the most famous example of a puzzle game is *Tetris* (Loguidice and Barton, 2009), which leverages the player's spatial recognition, pattern recognition, and strategic thinking. It is important to note that many puzzle games are either variations of earlier games or variations of existing nondigital games. Unlike other genres, puzzles games are not generally story based (although there are some notable exceptions, such as the 2007 game *Puzzle Quest*). Because puzzle games do not generally rely on advanced three-dimensional graphics, they are more common on mobile and handheld devices (such as the iPhone or the Nintendo DS) (Perez, 2009). An excellent example of this is the iPhone game *Trism*, a direct descendent of *Tetris*, which offers users near-infinite play length.

Adventure

Among the earliest games available for home computer systems (dating back as early as 1972 with Gregory Yob's game *Hunt the Wumpus*), adventure games are characterized by game play without reflex challenges or action (often referred to as "twitch" behavior) (Jones, 1997). Instead, adventure games are characterized by puzzle solving, exploration, and interaction with people and the environment. Game play is most often highly story driven. For this reason, adventure games are often closely associated with film and are considered to be cinematic in design, pacing, and storytelling. In addition, adventure games are usually nonconfrontational in nature. Noteworthy examples of adventure games include *Zork* (1980), *King's Quest* (1984), *Myst* (1993), and *The Longest Journey* (1999).

Early text-based adventure games (in which players were required to enter commands into the game via a keyboard) are sometimes referred to as "interactive fiction." An example of this is *GemStone IV* (1995).

Role-Playing Games

As a genre, role-playing games (RPGs) usually involve the player taking control of a character or characters while progressing through a relatively predetermined storyline. Core game mechanics in RPGs such as attribute-based character generation, experience points for character advancement, and statistically based combat all originate from traditional pen and paper, tabletop, role-playing games such as *Dungeons & Dragons* (Barton, 2008; Crigger, 2008).

As the name suggests, role-playing games require the player to take the role of their in-game character. Moreover, players often make an emotional investment in their character that is not seen in most other genres. Noteworthy RPGs include *Ultima* (1980), *Fallout* (1997), *Knights of the Old Republic* (2003), *Fable* (2004), and *World of Warcraft* (2005).

Action

All action games share a specific quality: game play rooted in hand-eye coordination and reaction time.[1] The vast majority of challenges found in action games are tests of the game player's physical

---

[1] Available from http://en.wikipedia.org/wiki/Action_game. Accessed June 2, 2009.

skill. While action games incorporate challenges such as puzzles, races, or object collection, these are simply game play mechanisms and in no way central to the genre.

Action games are easily the most common type of game—so common, in fact, that one might easily argue that they are a meta-genre populated by a series of distinct subgenres, such as platformers, first-person shooters, and third-person fighting games. Noteworthy action games include arcade titles such as *Space Invaders* (1978), platformers such as *Super Mario Brothers* (1985) or *Psychonauts* (2005), stealth action games such as *Assassin's Creed* (2007), fighting games such as *Street Fighter IV* (2009) or *Mortal Kombat* (1992), music-based games such as *Guitar Hero* (2005) or *Donkey Konga* (2003), or first-person shooters such as *Halo* (2001) or *Doom* (1993).

First-Person Shooters

While first-person shooters (FPSs) are generally thought of as a subgenre of action games, they are important and pervasive enough that it might be argued that they could easily stand as their own genre.

First-person shooters, as the name implies, put the player in the role of the protagonist, providing a first-person perspective with which to interact with the world. As action games, FPSs are usually firearm based and provide game play rooted in the player having swift hand-eye coordination and reaction time (Garmon, 2005). Classic FPSs include *Doom* (1993), *Quake* (1996), *Half-Life* (1998), *Unreal* (1999), *Halo* (2001), and *America's Army* (2002). FPS games appear to define the games industry for activists such as former lawyer Jack Thompson, who has publicly argued that these games are "murder simulators" and influence teenage players to become more violent (Leung, 2005), although connections between violent FPS games and violent behavior have been both supported and repudiated by the scientific community (Anderson and Dill, 2000; Wang et al., 2009). Perhaps the most innovative contribution of the FPS genre was the capability (via embedded editors) for players to create their own game "levels" (three-dimensional maps with objects and AI-driven opponents), which could then be distributed via the Internet and used for multiplayer matches between distributed opponents.[2] This predated the widespread acceptance of massively multiplayer online games (MMOGs) and helped form the foundation for the concept of user-generated content on the Internet today (Keiser, 2006; Jenkins, 2006). The phenomenon is epitomized by the game *CounterStrike* (1999), a user-created "mod" (modification) to the FPS *Half-Life* (1998), which was eventually offered commercially and has outsold the original commercial offering (Lister et al., 2009).

Strategy

The driving principle of strategy games is generally a measured and thoughtful management of resources (both human and natural) from a god-like perspective. Digital strategy games are most often conflict-based (e.g., military, social, economic) models of game play in which the player is pitted against either a single or multiple computer-controlled entity (Rollings and Adams, 2003). The strategy genre is usually divided into two subgenres: real-time strategy (RTS) and turn-based strategy (TBS).

In the case of RTS games, action is continuous, and players are required to make their decisions and actions within the fabric of a constantly changing game state. RTS game play is most often characterized by the acquisition of natural resources, the construction of a production infrastructure (factory, barracks, etc.), the research of technologies, and the production of units (troops, vehicles, etc.). The game play in

---

[2]Available at http://www.fpscreator.com/about.html. Accessed June 2, 2009.

many RTS games is characterized by military conflict, with the win conditions of the game being destruction of a human- or computer-controlled opponent (Adams, 2006). While there are several examples of RTS games designed specifically for home consoles, they are generally geared exclusively toward the PC, given that the RTS control interface is better suited to the "hot-key" nature of a keyboard compared to the limited control of a console controller. Noteworthy RTS games include *Dune 2* (1992), *Command and Conquer* (1995), *Starcraft* (1998), *World in Conflict* (2007), and *Halo Wars* (2009).

In the case of TBS games, game play is characterized by a sequential order of play in which each player (either human- or computer-controlled) is allotted a period in which they are free to analyze the game conditions and commit decisions to actions. After a player signals that he or she is finished with their actions, the play sequence cycles to the next player, who has the same opportunity to analyze the state of game play and commit decisions to actions. In this way, TBS games are similar to nondigital tabletop role-playing games, collectible card games, collectible miniature games, and designer board games. Noteworthy digital TBS games include the *Civilization* series, the *Heroes of Might and Magic* series, and the *Total War* series.

In social and historical model-based strategy games (both TBS and RTS), such as *Civilization*, *Sins of a Solar Empire*, or *Age of Empires*, game play is mediated by a "social engine" that simulates the interaction of complex social and economic variables. In certain cases, such as the *Civilization* franchise, the "social engine" is based on models of cultural evolution and change whose theoretical roots are found in 19th- and early 20th-century anthropology. These models, such as Elman Service's band/tribe/chiefdom state model (Service, 1963) or earlier unilinear cultural evolutionary models (e.g., Morgan's model formulated in 1877, Tylor's model formulated in 1871, the Durkheim model published in 1900), are considered to be over simplified and inaccurate by modern social anthropologists (Watrall, 2002a,b). As such these RTS games should not be considered accurate social-historical simulators.

"God" Games and Other Simulation Games

This genre is based on an underlying simulation of real-world tasks and is discussed in detail in the section "Political and Other Simulation Games" in Chapter 3.

Alternate-Reality Games

The genres explained thus far have been major historically active game genres. As technology expands and new types of games become possible, new genres begin to form. Similarly, some genres (e.g., interactive fiction) tend to depreciate over time, becoming more niche in their status. One such emergent genre gaining a lot of attention lately is alternate-reality games (ARGs; Jenkins, 2008). A description of ARGs shows how games are morphing into new forms that can permeate other types of emergent media online. In ARGs, game play reaches into player's lives via everyday technologies like Web sites, blogs, e-mail, and mobile devices, blurring the line between in-game and out-of-game experiences.[3] Stories flow from one platform to another, as players are challenged to discover bits and pieces of often cleverly hidden content, which contributes to the complete narrative picture. While ARGs began as marketing tools for products or services, they have evolved into a rich and compelling entertainment environment for telling engaging and immersive stories. ARGs are highly mediated experiences in which the game is facilitated, controlled, and paced by teams of "puppet masters" (Gosney, 2005; Rose, 2007).

---

[3]Available at http://www.argn.com/about/. Accessed June 1, 2009.

Technically ARGs are a form of MMOG, with individual ARGs attracting player bases numbering in the tens or hundreds of thousands and with a heavy slant toward online media.[4] However, online platforms for ARGs are used less as a narrow framework to deliver a tightly defined gaming experience than simply as a convenient, cheap, mass communication medium (Szulaorski, 2005; Edery and Mollick, 2008). While the typical MMOG uses a custom client (an application running on the player's home computer) that delivers and controls all content and interaction, ARGs use any application available on the Internet, such as e-mail or social network platforms, and potentially any Web site, as a mechanism to deliver a rich and compelling overall game experience to the player.

Despite widespread coverage and some marketing-oriented successes, ARGs have still been dogged by a niche status in the overall marketplace, and several commercial ARGs have flopped (e.g., Electronic Arts's *Majestic*, Mind Candy's *Perplex City*). However, the player behaviors associated with ARGs remain of great interest to large organizations seeking to foster new forms of collaboration, and groups like the Institute for the Future still actively pursue them in hopes of finding ways to utilize their design patterns for larger-scale problem solving.

Serious Games

"Serious games" is the title given to projects that utilize video game technologies and/or design techniques or that build complete games to address some need other than entertainment. Given the name, it might be easy to classify them as a genre of games, but actually serious games are more a field of activity than a distinct genre. In fact, most serious games are themselves members of a defined game genre (e.g., an RPG for teaching history or a "racing" game engine repurposed for driver education) with the prefix of "serious" added as a means to indicate the primary intent of the product's sponsor. More description and discussion of serious games as a field are given in Chapter 3.

Major Platforms

As provided earlier, a general definition for a platform refers to the combination of hardware and software that allows a game to operate. As with genres and player types, the platforms for which games are developed vary widely. In many ways these platforms directly impact both design and usage patterns.

Common gaming platforms include arcades, PCs, consoles, handheld devices, and mobile devices, as described below. Online games are not platforms, but often there are platforms that enable various types of online games. Given their nature to be separate from but also at times closely associated with platforms, online games are described here in terms of classification.

Arcade

As a platform or a locus for propagating game culture, arcades do not have the relevance they had in the mid-1980s, nor are arcade machines as ubiquitous as they were then. The arcades that remain are often housed in malls, amusement parks, and/or family entertainment centers/restaurants, such as the Dave & Busters chain. This is not entirely true worldwide, and even in North America there are some evolved aspects of arcades that still exist.

Outside North America, in Japan and parts of Asia the idea of the arcade as a physical "third place" filled with games still exists in the form of "PC Bangs" (*bang* is Korean for *room*), essentially Internet

---

[4]Available at http://www.unfiction.com/history, Accessed June 2, 2009.

cafes that exist predominantly to play MMOGs like *Lineage* and *World of Warcraft* or online team shooters like *CounterStrike*. These Internet game rooms are dominant throughout Asia, especially Korea and China, but exist in the Middle East as well.

As for the more traditional arcade that saw the debut of new titles and fancy cabinets, where titans such as *Sega, Atari, Namco*, and others slugged it out for billions in quarters, those days are long gone. The location-based entertainment business is now one more of amusement-level opportunities on one end and PC-centric Internet cafes or "bangs" on the other. The arcade as a platform is gone.

Personal Computers

While historically the dominant digital game platform, the PC has been eclipsed by the console as the predominant game platform since the late 1990s. The number of console games sold in 2007 was substantially higher than the number of PC games sold that year (Entertainment Software Association, 2008). However, the PC as a platform remains noteworthy for several reasons:

- It is still the most common platform for MMOGs.
- Since there is no control over the content developed for PCs (Windows, Mac, Linux, or otherwise), there is still a large amount of creativity and independent development on PCs. This importantly includes a lot of rhetorically oriented content (i.e., propaganda) and serious games-oriented content that would never be condoned on controlled platforms.
- The PC is still the dominant gateway to the Web and other evolved social networks; those offered by consoles are currently weak by comparison. While in some areas Internet-based phones (perhaps coupled with netbooks) will eclipse PCs, in Western countries the PC is still a critical source of connectivity for gamers. Even console households have PCs where players are able to join communities, find out news about their favorite games, and more. Web browsing on leading consoles is still a fraction of that done on PCs.
- The PC is still the dominant authoring environment for games and more importantly is still the means for creating most user-generated content for games (including those on consoles). The PC's interface and application power makes it the means by which gamers will create content, software hacks, helper applications, and so forth, for their favorite games. While some work has shown that consoles can play a larger role in allowing for user-generated content, it is likely that PCs working with consoles may still be the development platform for gamers for some time.

Console

Consoles represent the majority of the commercial digital games market in terms of both hardware and software. Video game consoles are appliance-like devices whose entire design serves its primary purpose to play digital games. Unlike PCs, consoles generally have fixed hardware and as such cannot be easily upgraded or altered. The current generation of consoles includes the Microsoft Xbox 360, the Sony PlayStation 3 (PS3), and the Nintendo Wii, shown in Figures E-1 and E-2. Traditionally, console games reside on removable media, such as cartridges or optical media. However, the most recent generation features an online distribution service that allows users to download games for a fee on to a form of nonvolatile storage, typically a hard disk or flash memory.

The current generation of consoles (e.g., Xbox 360, PS3, and Wii) also serve as platforms for social networks. Each features online services that support friend lists, text chat, voice chat, ranking, recommendation systems, and online identity generation and representation. While PC and Web-based offer-

FIGURE E-1 PlayStation 3 and Xbox consoles. SOURCE: See http://cybernetnews.com/do-you-know-the-capabilities-of-your-next-gen-game-console. Image courtesy of CyberNet News.

FIGURE E-2 Wii console with accelerometer-based control. SOURCE: Image courtesy of Nintendo of America, Inc.

ings like Facebook are more advanced in some respects, the console companies are working toward a seamless integrated social network and play experience. Xbox Live!, for example, shows the depth to which these new social services are being developed. Its "Party System" allows players to connect with up to eight friends and remain with those friends as the player moves from service to service. A player can seamlessly move from watching online Netflix movies together (if all members of the group have Netflix accounts) to playing cooperatively in a multiplayer game such as *Gears of War* or *Halo*. Sony and Nintendo also have multiplayer social systems, but they are less developed than those of Microsoft. PC systems such as *Valve's Steam* and *Kongregate* and systems that mesh with Facebook (which even the console companies may do soon) are also in the mix.

Looking at the distribution capabilities of digital and social services, consoles appear to be transitioning from game systems into full-featured entertainment platforms. However, they are doing this on gaming terms, providing tight experiences that eventually blur the lines between streamed audio/video, interactive experiences, social exchanges, and games. For all the advances made in digital distribution, a majority of retail revenue is still derived from sales of consumer packaged goods (Gaudiosi, 2009). The next generation of systems is likely to move to a more dominant digital distribution system for their offerings.

The question then is, What happens to the console as it exists now? Some companies like newly formed OnLive! (www.onlive.com) foresee a totally digital distribution system where game play is rendered on server farms and piped as video back to a small thin client that resides in the home, with a monthly subscription fee to a service. This cloud model debuted at the 2009 Game Developers Conference but was fairly criticized by skeptical industry developers. The future more likely holds a mix of cloud services and locally processed games. Thus, the console is moving from its more stoic one-off game system toward being a balanced blend of PC capabilities, cloud computing services, locally processed applications, and dynamic control interfaces (e.g., cameras, motion sensors) with some evolutionary capability, all in the name of dominating access to entertainment experiences at home. It is a titanic battle that shapes many aspects of where games will go as a result.[5]

Handheld

Handheld game devices are lightweight, portable, dedicated console units with integrated display, controls, and audio. The current generation of handheld game devices—the most popular of which are the Sony PlayStation Portable (PSP) and the Nintendo Dual-Screen (DS, or the recently released DSi)—feature integrated WiFi, with which users can connect to the Web, play wirelessly (either locally or over the Internet) with other players, or, in the case of the Nintendo DSi, purchase and download additional games.[6] As of February 2009, 50 million units of the Sony PlayStation Portable had been sold worldwide (Robinson, 2009), whereas 100 million units of the Nintendo DS had sold worldwide (Kelly and Wyman, 2009).

Handheld games are generally far cheaper than PC or console games. As such, they are designed to be accessible to a wider variety of users. Partially as a result of pricing and marketing (especially in the case of the Nintendo DS) and partially as a result of adoption patterns, handheld game devices have been broadly adopted and may not have the negative subcultural affiliations of console and PC games and gamers (Patsuris, 2004).

Such devices have sometimes created problems for publishers, which have found it difficult to market products with the same level of audience consistency they see on other systems. Despite these problems, Nintendo has shown that its DS device, coupled with more obtuse software offerings like *BrainAge* and puzzle games, can successfully attract an older demographic. This example provides evidence that as the population ages and video game developers get sharper about their software and hardware offerings, games can maintain a much higher level of popularity than once assumed.

Mobile

With robust and powerful platforms like the iPhone or Google's Android OS that marry ubiquitous connectivity with rich graphics, versatile mobile devices are seeing an amazing amount of growth in the game design industry (Wingfield and Lawton, 2008). However, prior to the iPhone's success, mobile, while growing, was a landscape of regressed innovation and tough markets. There were many different types of mobile devices that required heavy porting of games across many different specifications to achieve a reasonable level of market opportunity. Furthermore, since games had to be downloaded directly from the local telecommunications operator, consumers faced enormous gate-keeping restrictions. This setup essentially stunted mobile game opportunities for some time. Only as the new genera-

---

[5]For additional information on cloud computing, see the following Web site: http://www.supercomp.de/isc09/Program/At-a-Glance/Cloud-Computing-HPC-Synergy-or-Competition. Last accessed October 14, 2009.

[6]Available at http://en.wikipedia.org/wiki/Nintendo_DSi. Accessed June 2, 2009.

tion of mobile smart devices, such as the iPhone, come about—where applications are not as controlled by the telecommunications services partners and the system is much more standardized, robust, and installed in large homogenized numbers—has mobile gaming use begun to explode.

Most games available on mobile devices tend to be casual in nature in that they are designed specifically to be played in small chunks with little investment in time. However, as devices and services get more sophisticated and specific types of devices, such as the iPhone, see a higher degree of market penetration, there is beginning to be a rise in other game genres. One of the most recent examples of this is *Watchmen: Justice Is Coming*, a massively multiplayer online role-playing game (MMORPG) for the iPhone designed to let players collectively experience the world of Alan Moore's *Watchmen* comic.

The iPhone is a critical glimpse at the future of mobile phone games. It is essentially a robust computer, game console, and phone with much more lenient publishing rules than any other device in its category. As it and similar offerings from other providers evolve into the market, lower prices, and hit wider global audiences, it is evident that computing power for games and other applications will be available to people in the coming years even if they never own a true PC or game console. (See "Mobile Games Platforms" in Chapter 3 for further analysis.)

Online

The term "online games" covers an amalgamation of game types and processes. The ambiguity of the term leaves that audience confused or assuming online games only means "massively" multiplayer online games,[7] a subset of a larger set of games played online.

Online games consist of the following broad categories:

- **Single-player online games:** These games are accessed via an online connection usually through a Web browser. These games often are written in Java or Flash and are accessed by visiting a Web site that contains the game.
- **Multiplayer Online Games:** These are games in which players play against other opponents through online connections. Such games can feature up to 128 players per match, although these are usually 2 to 32 players, in teams battling against one another. These games once existed offline and were played over local area networks, but now they are predominantly played online.
- **Massively Multiplayer Online Games (MMOGs):** These games feature an ability to serve thousands or even millions of players. Such games are often also persistent worlds where the game state is consistently evolving over time whether a given player participates or not.
- **Social Network Games:** These games are a relatively new form of online game popularized by games embedded on sites like Facebook. These games use the social network connections a user has specified to foster various forms of peer-to-peer game play. Such work is still in its infancy, but several games are played by millions of users and often can generate a lot of messaging to go alongside the game play.

---

[7]The term "massively multiplayer online role-playing games" is slightly misleading. It is used colloquially to refer to all online persistent virtual worlds. However, it is more often used to refer to a genre of massively multiplayer online games, specifically fantasy-based RPGs such as *World of Warcraft*, *Everquest*, *Dungeons & Dragons Online*, or *Warhammer Online*. In recent years, with the increasing success of MMORPGs, new online multiplayer genres have begun to emerge, such as MMOFPSs (massively multiplayer online first person shooters) and MMORTS (massively multiplayer online real-time strategy) (Ryan, 2007). Most people call these MMOs or MMOGs (massively multiplayer online games) and do not use the longer, more specific acronyms.

The fastest-growing online MMOGs at this point in time are those connected with social media sites, such as Facebook. The game company Zynga produces many of these hits, including *Mafia Wars*, *Yoville*, and several others, that together earned the company $50 million in microtransaction revenue in 2008 (Chowdhry, 2009). The integration of these games with Facebook means that Zynga is able to ride on Facebook's ascent and growing media presence. With Facebook's ability to rally together like-minded groups for social, intellectual, and personal causes, Zynga is well positioned to create and sell targeted games.

MMOGs online are also somewhat confusingly named. Often neophytes will think that these games have an ability to house all players in the same shared space or even game. In fact, only a few games and virtual worlds achieve such a "full single world" state (namely *Second Life* and *Eve Online*), because technical limitations of hardware would make it nearly impossible to do this, let alone an interface for guiding people around a virtual world where everyone crowds into the same arena for battle. Instead, most MMOGs are segmented into mirrored versions of the same geographies, and then there are limits to how many users can log into the server at a single time. Thus, a game such as *World of Warcraft* with millions of players spreads them across hundreds if not thousands of servers, each supporting a fraction of the total subscriber base expected online at any specific instance. Furthermore, segments of the games (called instances) reduce the number of players to an area down to 20 to 50 people at a time to enhance the response time and minimize player confusion. From a collaboration standpoint, observations, and so forth, a player on Server 1 might never come into contact with a player on Server 22 in a game like *World of Warcraft*. Despite these limiting factors, such games are still fairly massive in scope and are dynamic cauldrons of social behavior that can only erupt from such a large number of players in the same communal space.

Player Types

Whether committed or casual gamers, players of all stripes find themselves shaped by the games they play and the ways they play them. As with genres, players exhibit a great deal of diversity as to their motivations, play styles, preferences, and characteristics. One widely recognized and respected method of player classification is known as the Bartle Test of Gamer Psychology (Bartle, 1996). Developed by Richard Bartle, one of the original developers for the first multiplayer online game (or MUD, from *Multi-User Dungeon*), the system categorizes player motivation as a given percentage of four categories commonly referred to as Bartle's player types: achievers, explorers, socializers, and killers. While these categories were developed based on Bartle's observation of MUD players, they are equally applicable to nonmultiplayer games.

Achievers

Achievers view point-based reward gathering as the primary motivation for the game. Other in-game activities such as socializing are either subservient or a means to accomplish goals linked to point gathering.

Explorers

Explorers delight in testing the boundaries of the game and having its inner workings revealed to them. Explorers revel in finding content or features (either intentional or unintentional on the part of the developers) not widely or commonly known to other players (e.g., hidden treasures in adventure games or unreported key combinations leading to new outputs in action games).

Socializers

Socializers, most common for multiplayer games, play to make social connections. They strive to know fellow players better by using in-game communication (chat, voice, etc.) and often transfer that communication out of game. Socializers either seek out or find in-game social groups (guilds) and often go out of their way to help other players in game, especially those who are beginners or at a lower level than the socializer. Socialization in gaming is further described in Chapter 3.

Killers

With a name chosen as much for its effect as its characteristics, killers do not necessarily relate to actually killing other players unless such an act is a metaphor for defeating opponents in a game. For most players of the "killer" type, the joy of game playing comes from a friendly competitive spirit. These players are in it for the sport, strategically pitting their abilities against those of an opponent (either computer-controlled or a fellow player). For others, especially in a massively multiplayer setting, game play motivation is more about power and the ability to dominate other players, especially those who are less powerful. They love being someone feared or, even better, someone to be attacked immediately by other, more "ethical" players in the game, or "killed on sight" in the common gaming vernacular.

Technical Design Patterns

Genre categories are not the only way to break down the types of games or associated design patterns. Video games by definition have associated technology that underlies them and provides the means to access and play the game. Technical classification of games can be defined in terms of the platform the game is made for (e.g., Windows PC, PlayStation Portable, or a game console such as the Nintendo Wii) or the specific technologies used to author it or play it back on a specific piece of computer hardware. What can be confusing here is that some platforms are defined by their associated hardware (e.g., PlayStation), while other platforms are defined by the host operating system (e.g., Mac OS X or Windows Vista) or by the virtual machine that is used to play it back regardless of the associated operating system (OS) or hardware such as Flash, Java, or asynchronous JavaScript and XML (AJAX). Hardware-oriented platforms like PlayStation are often combinations of hardware, OS, and some associated technologies, while specific OS or virtual machine labels may be more exclusive. For example, the Wii platform is the combination of specific hardware and a very specific OS designed by Nintendo, but it also includes a Web browser with Flash virtual machine support. So a well-designed Flash game is playable on the Wii platform, just as it is accessible to people with Macs and PCs with Flash support.

Being able to identify and parse technical design patterns and platforms is relevant in order to denote the capability of the games that utilize them and thus gain an immediate sense of the targeted audience, features offered, and other strategically important factors such as their business model. This is made even more important by the fact that today games are becoming more and more ubiquitous across many modalities and, as such, many types of platforms are seeing robust development. Flash games are being played millions of times daily. Games developed specifically for Facebook's social application programming interfaces are growing frantically, while *RuneScape*, a Java-based MMOG, has over 10 million player accounts. The committee expects that in the next 20 years there will be even more platforms and technologies enabling game play.

THE BUSINESS OF GAMES

Industry Structure

The modern games industry has a number of key players, each of whose role is described below.

Console Developers: Driving Platform Innovation

Many video games are played on dedicated devices designed specifically for game play, known as consoles. In the modern hardware manufacturing game, hardware platform development is expensive and time consuming, with the typical hardware platform having a 2- to 3-year development cycle and a 5- to 10-year market lifetime (with an upgrade or two during that cycle for the serious player; Ivan, 2009).

The latest versions of the three main console platforms have achieved impressive worldwide installed bases. The Wii, which focuses more on casual games, has achieved a meaningful footprint in the market. Considered more accessible, less intimidating, and more "natural" for nongamers to play with its innovative accelerometer-based control without the need for complex key combinations, the platform has broadened the user base for console games (Morrison, 2008). Figure E-3 shows hardware sold through May 2009 for three major consoles.

In addition to their strong hardware footprints, these hardware manufacturers have built ecosystems that allow massive numbers of software developers to build large franchises, as illustrated by the growth in software sales shown in Figure E-4.

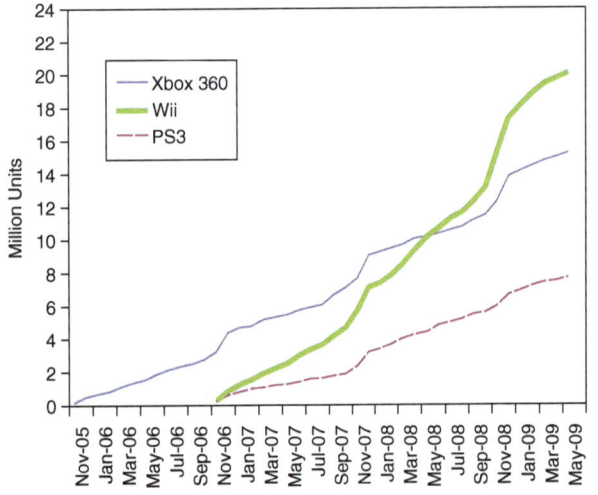

FIGURE E-3 Xbox 360, Wii, and PS3—hardware units, lifetime sales through May 2009. SOURCE: Data from NPD Group, Deutsche Bank.

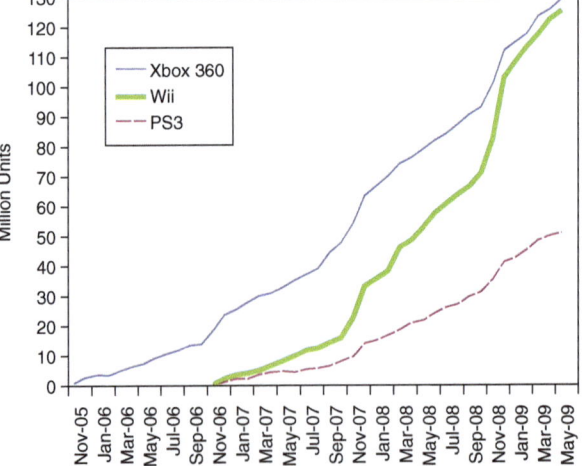

FIGURE E-4 Xbox 360, Wii, and PS3—software units, lifetime sales through May 2009. SOURCE: Data from NPD Group, Deutsche Bank.

Game Developers: The Engine of the Games Industry

Game developers are the lifeblood of the games industry. Game development teams bring new intellectual property to life by combining technical and creative talent to design and build games they believe will succeed in the marketplace. These teams can be organized in many different ways. Independent game developers (also known as "indies") are generally small teams of people focused on creating a breakout title. These developers create relationships with other third parties to get wider distribution and marketing support for the games they create. Because indie developers are often financially constrained, they often ally themselves with publishers to get financial and go-to-market support for new games.

Many large publishers also have development studios of their own. Publisher studios tend to be larger than independent game development studios with a focus on the creation of a particular game title or franchise. There is additionally a submarket of component and content developers that support some studios, much like the support organizations found in other major industries.

Publishers: Key Marketing and Distribution Conduits

Publishers are a key part of the games industry value chain. They are responsible for bringing developers' intellectual property (IP) to market. Publishers support game developers in a number of ways: by providing financing for new game development, by maintaining relationships with online and retail distribution outlets, and by facilitating marketing support for new game launches. By way of analogy, game publishers are similar to movie studios in terms of the marketing and distribution roles they play in the industry.

The two dominant independent (i.e., not tied to any specific hardware platform) publishers in the United States are Electronic Arts and Activision Blizzard; French-based Ubisoft is the other major publisher by marketshare (and is 20 percent owned by Electronic Arts); other major publishers include Japanese-based Namco/Bandai, Konami, Square|Enix, and Capcom. Sony, Microsoft, and Nintendo are major platform holders and also have significant in-house publishing arms focused on developing and marketing games designed for play on their respective platforms. As the games industry shifts from retail distribution to online distribution, the role of the publisher continues to morph and consolidate. As the distribution for games shifts online, other groups may eventually rise to play significant publishing-styled roles. Already, Time-Warner and Viacom have become major owners of game studios and are creating new publishing groups for games such as *Rock Band* and *Lego Star Wars*.

Peripherals and Accessories

In addition to game consoles and software, there is a robust market for game peripherals and accessories. Peripherals and accessories are generally designed to improve or augment game play by providing a more comfortable game-playing experience. On the console side, peripherals and accessories can add network connectivity, voice-over-IP (Internet protocol) technology, advanced controls, and navigation. For many of these accessories (including the iPhone), just like their software counterparts the producers must pay a licensing fee per unit to gain the right to market compatible hardware. This is yet another reason ownership of a successful console platform can be a large money maker.

While at first peripherals were only a sideshow for the industry, the recent rise of simulative rhythm games such as *Rock Band*, *Guitar Hero*, *DJ Hero*, *DanceDanceRevolution*, and more has seen the role of controllers and peripherals become much bigger in scope. Games are increasingly seen as performance based, and controllers are an essential contributor in this process.

Games Industry Business Models

As the games industry continues to mature, the ways in which developers generate revenue continues to evolve. There are three dominant business models in the games industry—packaged consumer goods, subscription software and services, and microtransaction-based—with some players employing hybrid strategies that combine a number of the discrete models described below.

Packaged Software

Historically, the games business has been driven through the sales of packaged products. Such a model was natural, especially given that the industry's highest-level employees have traditionally come from packaged consumer goods industries. The majority of game sales occur through traditional retailers, with the consumer purchasing a physical copy of the game for use on a home computer or game console. Publishers are responsible for securing retail distribution and promotion for game titles, and large electronics retailers have been the most efficient way to get a new game in front of a mass audience (Riley, 2008).

The dominance of retail distribution continues to decline as the Internet and digital delivery become more important mechanisms for getting games to consumers (Macrovision Corporation, 2003). This is especially true in many non-Western markets where IP laws are regularly curtailed and the most dependable revenue streams for game sales are through online-distributed games accessed via a validated subscription. In Korea, for example, online games are already the norm rather than the exception. Business models are expected to change over time at a slower pace in the West, but the digital distribution model is widely expected to dominate in the next decade.

Subscription Software and Services

With the advent of broadband connectivity has come a second approach to package and price products for end users. Whereas the packaged software model involves one-time transactions, the subscription model allows a user to pay a small amount on a monthly basis for ongoing access to a game hosted online. The publisher or game developer takes responsibility for hosting and maintaining the game via a Web site or downloadable client software application.

Subscription software and services (such as the Xbox Live! service) have been one of the fastest-growing submarkets in the games industry. Subscription software has the additional benefit of providing the publisher with a recurring revenue stream to support future development. Subscription games have also allowed games to reach a wider audience as they do not require proximity to a retail location to acquire a copy of the game and most do not require a specialized game console.

Free-to-Play Games with Microtransactions

The free-to-play business model, which generates revenue predominantly through the collection of microtransaction purchases of in-game items, is discussed in depth in the corresponding section in Chapter 3.

REFERENCES

Adams, Dan. 2006. The state of the RTS. *IGN Entertainment*, April 7. Available at http://pc.ign.com/articles/700/700747p1.html. Accessed June 1, 2009.

Anderson, Craig A., and Karen E. Dill. 2000. Video games and aggressive thoughts, feelings, and behavior in the laboratory and in life. *Journal of Personality and Social Psychology* 78(4):772-790.

Bartle, Richard. 1996. Hearts, clubs, diamonds, spades: Players who suit MUDs. *The Journal of Virtual Environments* 1(1). Available at http://www.brandeis.edu/pubs/jove/HTML/v1/bartle.html. Accessed June 23, 2009.

Barton, M. 2008. *Dungeons and Desktops: The History of Computer Role-Playing Games*. Wellesley, MA: A. K. Peters, Ltd.

Chowdhry, Amit. 2009. Social network gaming company Zynga pulling in $100 million this year. *Pulse2*, May 3. Available at http://pulse2.com/category/zynga/. Accessed July 9, 2009.

Crigger, Lara. 2008. Chasing D&D: A history of RPGs. *1UP.com*. Available at http://www.1up.com/do/feature?pager.offset=0&cId=3168091. Accessed June 15, 2009.

Edery, David, and Ethan Mollick. 2008. *Changing the Game: How Video Games Are Transforming the Future of Business*. Upper Saddle River, NJ: FT Press.

Entertainment Software Association. 2008. *Essential Facts About the Computer and Video Game Industry*. Washington, DC: Entertainment Software Association. Available at http://www.theesa.com/facts/pdfs/ESA_EF_2008.pdf. Accessed January 13, 2010.

Garmon, J. 2005. Geek trivia: First shots fired. *Tech Republic*, May 24. Available at http://articles.techrepublic.com.com/5100-10878_11-5710539.html. Accessed June 1, 2009.

Gaudiosi, John. 2009. Digital downloads spell end for videogame stores? *Reuters*, June 11. Available at http://www.reuters.com/article/internetNews/idUSTRE5596CX20090611. Accessed July 22, 2009.

Gosney, John W. 2005 *Beyond Reality: A Guide to Alternate Reality Gaming*. Boston: Thomson.

Ivan, Tom. 2009. Sony: Rivals looking at a five year lifecycle. *Edge Online*, June 10. Available at http://www.edge-online.com/news/sony-rivals-"looking-at-a-five-year-lifecycle." Accessed January 13, 2010.

Jenkins, Henry. 2006. *Fans, Bloggers, and Gamers: Exploring Participatory Culture*. New York: NYU Press.

Jenkins, Henry. 2008. *Convergence Culture: Where Old and New Media Collide*. New York: NYU Press.

Jones, Marshall G. 1997. Learning to play; playing to learn: Lessons learned from computer games. Conference of the Association for Educational Communications and Technology. Available at http://www2.gsu.edu/~wwwitr/docs/mjgames/index.html. Accessed July 2, 2009.

Keiser, J. 2006. Playing for change: How video game players have become pioneers. *1UP.com*, August 31. Available at http://www.1up.com/do/feature?pager.offset=0&cId=3153059. Accessed June 15, 2009.

Kelly, Andrew, and Michele Wyman. 2009. Nintendo ships 100 millionth portable Nintendo DS system. *Business Wire*, March 11. Available at http://www.businesswire.com/news/home/20090311006007/en. Accessed May 7, 2009.

Leung, Rebecca. 2005. Can a video game lead to murder? Ed Bradley interview of Jack Thompson. *60 Minutes*, June 19. Available at http://www.cbsnews.com/stories/2005/06/17/60minutes/main702599.shtml. Accessed June 11, 2009.

Lister, Martin, Jon Dovey, Seth Giddings, Iain Grant, and Kieran Kelly. 2009. *New Media: A Critical Introduction* (2nd Ed.). New York: Routledge.

Loguidice, Bill, and Matt Barton. 2009. *Vintage Games: An Insider Look at the History of Grand Theft Auto, Super Mario, and the Most Influential Games of All Time*. St. Louis, MO: Focal Press.

Macrovision Corporation. 2003. Trymedia Systems releases ActiveMARK 4.0 and software development kit. *Trymedia Systems*, February 24. Available at http://www.trymedia.com/corporate/press/20030224.shtml. Accessed January 13, 2010.

Morrison, Brice. 2008. Opinion: Two years in—The Wii's successes. *Gamasutra.com*, October 21. Available at http://www.gamasutra.com/php-bin/news_index.php?story=20676. Accessed June 12, 2009.

Patsuris, Penelope. 2004. Sony PSP vs. Nintendo DS. *Forbes.com*, June 7. Available at http://www.forbes.com/2004/06/07/cx_pp_0607mondaymatchup.html. Accessed June 15, 2009.

Perez, Marin. 2009. Games rule iPhone app downloads. *Information Week*, April 8. Available at http://www.informationweek.com/news/personal_tech/iphone/showArticle.jhtml?articleID=216403536. Accessed June 2, 2009.

Riley, David M. 2008. Extreme gamers spend an average of 45 hours per week playing video games. *The NPD Group, Inc.*, August 11. Available at http://www.npd.com/press/releases/press_080811.html. Accessed January 13, 2010.

Robinson, Andy. 2009. PSP: 50 million sold. *Computer and Video Games*, February 13. Available at http://www.computerandvideogames.com/article.php?id=208211%3fcid&skip=yes. Accessed May 7, 2009.

Rollings, Andrew, and Ernest Adams. 2003. *Andrew Rollings and Ernest Adams on Game Design*. Indianapolis, IN: New Riders Games.

Rose, Frank. 2007. Secret Websites, coded messages: The new world of immersive games. *Wired* 16(01). Available at http://www.wired.com/entertainment/music/magazine/16-01/ff_args. Accessed June 23, 2009.

Ryan, Leon. 2007. Beyond the looking glass of MMOG's. *GameAxis Unwired,* pp. 27-31. May. Available at http://books.google.ca/books?id=vOoDAAAAMBAJ&pg=PA29&dq=%22MMOFPS%22&client=firefox-a#PPA27,M1. Accessed March 15, 2009.

Service, Elman Rogers. 1963. *Primitive Social Organization: An Evolutionary Perspective.* New York: Random House.

Szulaorski, Dave. 2005. *This Is Not a Game: A Guide to Alternate Reality Gaming.* Santa Barbara, CA: eXe Active Media Group.

Wang, Yang, Vincent P. Mathews, Andrew J. Kalnin, Kristine M. Mosier, David W. Dunn, Andrew J. Saykin, and William G. Kronenberger. 2009. Short term exposure to a violent video game induces changes in frontolimbic circuitry in adolescents. *Brain Imaging and Behavior* 3(1):38-50.

Watrall, Ethan. 2002a. Interactive entertainment as public archaeology. *Society for American Archaeology Archaeological Record* 2(2):37-39.

Watrall, Ethan. 2002b. Digital pharaoh: Archaeology, public education, and interactive entertainment. *Journal of Public Archaeology* 2:163-169.

Wingfield, N., and C. Lawton. 2008. Apple's iPhone faces off with the game champs. *Wall Street Journal*, November 12, p. B1. Available at http://online.wsj.com/article_email/SB122644912858819085-lMyQjAxMDI4MjE2MjQxNDI5Wj.html. Accessed June 2, 2009.